Dmitry Shestopalov
Larissa Golubeva
Edward Cloutis

Optical effects of space weathering in the main belt of asteroids

AF153245

Dmitry Shestopalov
Larissa Golubeva
Edward Cloutis

Optical effects of space weathering in the main belt of asteroids

LAP LAMBERT Academic Publishing

Impressum / Выходные данные

Bibliografische Information der Deutschen Nationalbibliothek: Die Deutsche Nationalbibliothek verzeichnet diese Publikation in der Deutschen Nationalbibliografie; detaillierte bibliografische Daten sind im Internet über http://dnb.d-nb.de abrufbar.

Alle in diesem Buch genannten Marken und Produktnamen unterliegen warenzeichen-, marken- oder patentrechtlichem Schutz bzw. sind Warenzeichen oder eingetragene Warenzeichen der jeweiligen Inhaber. Die Wiedergabe von Marken, Produktnamen, Gebrauchsnamen, Handelsnamen, Warenbezeichnungen u.s.w. in diesem Werk berechtigt auch ohne besondere Kennzeichnung nicht zu der Annahme, dass solche Namen im Sinne der Warenzeichen- und Markenschutzgesetzgebung als frei zu betrachten wären und daher von jedermann benutzt werden dürften.

Библиографическая информация, изданная Немецкой Национальной Библиотекой. Немецкая Национальная Библиотека включает данную публикацию в Немецкий Книжный Каталог; с подробными библиографическими данными можно ознакомиться в Интернете по адресу http://dnb.d-nb.de.

Любые названия марок и брендов, упомянутые в этой книге, принадлежат торговой марке, бренду или запатентованы и являются брендами соответствующих правообладателей. Использование названий брендов, названий товаров, торговых марок, описаний товаров, общих имён, и т.д. даже без точного упоминания в этой работе не является основанием того, что данные названия можно считать незарегистрированными под каким-либо брендом и не защищены законом о брендах и их можно использовать всем без ограничений.

Coverbild / Изображение на обложке предоставлено: www.ingimage.com

Verlag / Издатель:
LAP LAMBERT Academic Publishing
ist ein Imprint der / является торговой маркой
OmniScriptum GmbH & Co. KG
Heinrich-Böcking-Str. 6-8, 66121 Saarbrücken, Deutschland / Германия
Email / электронная почта: info@lap-publishing.com

Herstellung: siehe letzte Seite /
Напечатано: см. последнюю страницу
ISBN: 978-3-659-68113-4

Contents

Preface.. 2

1. Introduction.. 3

2. Optical maturity degree of asteroids and the Moon.. 9

3. Timescales of optical maturation of undisturbed surface: data of laboratory
   experiments.. 14

   3.1. Solar wind as a source of surface maturation.. 14

   3.2 Microparticle bombardment as a source of surface maturation.................. 16

   3.3. Summary... 17

4. Collisions between asteroids... 19

   4.1. The number and frequency of collisions... 19

   4.2. Craters and crater ejecta... 22

   4.3. Seismic activity of asteroids and regolith motions...................................... 29

   4.4. Short and long space weathering timescales... 35

5. Discussion.. 38

6. Conclusion.. 41

References 42

PREFACE

A supposition that the spectra of S asteroids are the modified spectra of ordinary chondrites due to space weathering effects results in the following paradox that needs to be clarified: The spectral modification process of main belt asteroids apparently terminates with S-type spectra on a timescale of ~ 10^5 yr passes and does not proceed to "completion"; i.e., lunar-type spectra over the next ~ 10^6 yr and beyond. This is especially puzzling since both these time intervals are much shorter than the collisional lifetime (~10^9 yr) of asteroids larger than about 10 km in size.

What is happening on asteroid surfaces on the space weathering timescale of ~ 10^5 yr, i.e., in the time needed to make any "chondritic" spectrum similar to the S-type? The expected time of impact onto asteroids turns out to vary from some hundred days to about ten hours depending on their size. Therefore, asteroid surfaces cannot be considered as undisturbed on this weathering timescale. We will demonstrate that impact activated motions of regolith particles hamper the progress of optical maturation of the asteroid surfaces.

The difference in the degree of optical maturation of the Moon and asteroids results from the unique properties of their regoliths. The uppermost layer of the lunar surface survives for ~ 10^7 yr before burial by another layer, so that lunar regolith particles have time to accumulate detectable amount of reduced iron and proceed to lunar-type maturation. Unlike the Moon, asteroid regoliths are characterized by higher particle mobility due to the high frequency of non-catastrophic collisions with projectiles and low gravity of asteroid targets.

This research utilizes spectra of meteorites measured by M. J. Gaffey, T. Hiroi, and R. P. Binzel available at NASA RELAB at Brown University (http://www.planetary.brown. edu/relab/) as well as spectral data on S asteroids from the catalogues ECAS, SMASS, 52 colors, Fieber-Beyer IRTF Mainbelt Asteroid Spectra, and IRTF Near-IR Spectroscopy of Asteroids available at NASA PDS (http://pdssbn.astro.umd.edu/). We thank all investigators for their spectral measurements used in our work.

1. Introduction

Space weathering is a catch-all term that encompasses physical, chemical, and morphological alterations of a planetary surface not covered by an atmosphere under the action of exogenous factors such as meteorite bombardment, temperature variations, solar wind particles, and galactic cosmic rays.

Solar wind sputtering and micrometeorite impact vaporization generate thin vapor-deposited coatings containing submicroscopic spherules of reduced iron (Fe^0) on regolith particles. In spite of the trace concentration of Fe^0 in regolith particles, nanometer-size iron particles lead to optical maturation of the lunar surface, i.e. to a decrease in its albedo and reddening of spectral slope. The latter is a brief summary of more than twenty years of research carried out by B. Hapke and his colleagues to find an answer to the question of why the lunar surface material is so dark and red, and why lunar reflectance spectra contain no strong absorption bands (Hapke, 2001 and references therein). Their finding was verified by transmission electron microscope images, which provided direct visual proof of the Fe^0-rich coatings on lunar regoliths particles (Keller and McKay, 1997; Wentworth et al., 1999; Pieters et al., 2000).

The mathematical theories developed by Shkuratov et al., (1999) and Hapke (2001) quantitatively describe the optical effects of the reduced iron particles, and have been applied to laboratory experiments simulating space weathering (Hapke, 2001; Shestopalov and Sasaki, 2003; Starukhina and Shkuratov, 2011; Lucey and Riner, 2011) and to the regoliths of atmosphereless cosmic bodies (e.g., Shkuratov et al., 1999; Starukhina and Shkuratov, 2001; Shestopalov and Golubeva, 2004). Recently Noble et al. (2007) found that the VNIR reflectance spectra of laboratory analogs imitating planetary soils depend significantly not only on concentration of the reduced iron in various samples but also on the size of the iron particles. Larger iron particles (>~ 200 nm) cause an overall decrease of spectral albedo and only modest modification of spectral curves while the small iron particles (<~ 40 nm) are responsible for darkening of the surface and its strong spectral reddening. A few years later Starukhina and Shkuratov (2011) and Lucey and Riner (2011) calculated this grain size effect mathematically using Mie theory together with the scattering theories referred to above. Hörz et al. (2005) demonstrated experimentally that impacts at ~ 5 km/s lead to melting of ordinary chondrite regolith and to

4

dissemination of pre-existing metals and sulfides. Such metal droplets (< 100 nm in size), disseminated throughout silicate glass, are supposed to produce space weathering effect on S-type asteroids if they, of course, have chondritic composition. Unfortunately, the absence of spectral measurements of altered chondritic samples in this investigation does not allow us to estimate this optical effect.

There is no doubt that space weathering is a process widespread in the Solar System but the intensity of the process and, consequently, its optical effect depend on the heliocentric distance, and the chemical and physical properties of planetary surfaces. We consider the matter of space weathering in the asteroid belt by the example of S asteroids since the optical maturation of silicate-bearing surfaces is studied sufficiently well. It can be expected that the Dawn mission to Ceres will help clarify the characteristics of space weathering on primitive asteroids containing ices and complex organic material.

The conclusion that asteroid surfaces are optically less mature than lunar soil was made at the end of the last century (Matson et al., 1977; Golubeva et al., 1980) and has been repeatedly corroborated (Pieters et al., 2000; Hapke, 2001; Golubeva and Shestopalov, 2003; Noble et al., 2007). The simplest evidence of this is the fact that weak absorption bands are seen in visible wavelength region of S-asteroid spectra whereas these bands have not yet been found in lunar spectra (e.g., Shestopalov et al., 1991; Hiroi et al., 1996). For example, a narrow spin-forbidden band of iron oxide near 505 nm is detected in spectra of dozens of asteroids of various optical types (Cloutis et al., 2010).

In spite of the low degree of weathering in the asteroid belt, this process could "confuse the cards" when interpreting reflectance spectra of asteroids. Space weathering is often considered as the cornerstone required to resolve the so-called "S-type conundrum", which consists of the apparent scarcity of ordinary chondrite parent bodies among S asteroids (Chapman, 1996, 2004; Pieters et al., 2000; Hapke, 2001). Laboratory experiments simulating micrometeorite impact vaporization and solar wind sputtering show that the spectra of ordinary chondrites and olivine-bearing samples do become similar in general to S- and A-asteroid spectra, respectively (Moroz et al., 1996; Yamada et al., 1999; Sasaki et al., 2001; Strazzulla et al., 2005; Brunetto et al., 2006a, 2006b; Dukes et al. 1999; Loeffler et al., 2008, 2009). Based on these experimental results, most researchers tend to conclude that most S asteroids have chondritic composition, which has been modified by space weathering. However, Gaffey (2010) notes that diagnostic mineralogical parameters, such as absorption band centers and band area ratios that allow mineralogical characterizations of asteroid mafic rocks, are not appreciably affected by asteroid-style space weathering. Since the spectral diagnostic parameters – as is evident from these experiments – are essentially unaffected by laser irradiation or ion beam treatment, Gaffey (2010) deduces that the alteration process fails to explain mineralogically important differences in spectra of ordinary chondrites and S asteroids and so their compositions appear to be different from those of ordinary chondrite meteorites. In addition, an increase in the spectral slope with increasing phase angles was found for near-Earth Q-, and S-type asteroids, as well as ordinary chondrites (Sanchez et al., 2012). The authors stress that the increase in spectral slope caused by phase reddening is comparable with the degree of space weathering and should be considered when studying space weathering effects on spectra of near-Earth asteroids.

The theory developed by Shkuratov et al. (1999) was used in our work (Shestopalov and Golubeva, 2004) to simulate optical maturation of ordinary chondrite surfaces. We investigated the variations of spectral characteristics of ordinary chondrites depending on the size of host material particles and volume concentration of fine-grained Fe^0 and found no matches with the same spectral characteristics as S-type asteroids. Thus we inferred that the spectral differences in question are caused not only by a weathering process but also systematic differences in the material compositions of S asteroids and ordinary chondrites.

The efforts of some investigators have been focused on estimation of a space weathering rate in the asteroid belt (Lazzarin et al,. 2006; Marchi et al., 2006a, 2006b; Paolicchi et al., 2009, 2007; Nesvorný et al., 2005; Willman et al., 2008; Willman and Jedicke, 2011). A principal motivating force of these studies was to find a rational explanation of the paucity of spectroscopic analogues of ordinary chondrite meteorites among silicate-rich S-type asteroids. If space weathering does take place in the main belt, then spectra of relatively young S asteroids (small near-Earth objects or the members of dynamically young families) should differ from those of relatively old S asteroids (large objects of the main belt or the members of dynamically old families). A spectral slope in the vis-NIR range was chosen as a space weathering index, which was supposed to be directly correlated with the different relative age of asteroids. Despite the fact that this choice cannot be recognized as most suitable for surface maturity characterization (Gaffey, 2010), the final result of the investigations was rather contrary to expectations (Paolicchi et al., 2009; Willman and Jedicke, 2011). It turned out that an astrophysical timescale (i.e., the time necessary to change spectra of, say, small Q-type asteroids to typical spectra of large S-type asteroids) is greater by some orders of magnitude than space weathering timescales estimated from laboratory experiments (e.g., Hapke, 2001; Loeffler et al., 2009; Brunetto and Strazzulla, 2005). As some investigators improved their age-color-size theory for asteroids, the value of asteroid weathering timescale varied directly with the date of publication: 570 ± 220 Myr (Willman et al., 2008), 960 ± 160 Myr (Willman et al., 2010), 2050 ± 80 Myr (Willman and Jedicke, 2011). Later on, the authors specified more correctly a lower bound of the weathering timescale of $\sim (5.1 - 7.9) \times 10^8$ yr (Willman and Jedicke, 2012). These investigations imply that the space weathering rate in the asteroid belt is very low or, equivalently, that the degree of weathering of asteroid surfaces is low. Moreover, Vernazza et al. (2009) deduce: *"The rapid color change that we find implies that color trends seen among asteroids are most probably due to compositional or surface-particle-size properties, rather than to different relative ages."* This conclusion confirms our result concerning the S-asteroid compositional trends with heliocentric and perihelion distances (Golubeva and Shestopalov, 1992; 2002). In turn, these compositional trends support the idea expressed by McFadden et al., (1985) and Nesvorný et al., (2009) that the fragments of a disrupted S asteroid could be transformed to the Earth-crossing orbits via 5:2 resonance with Jupiter at 2.823 AU.

A polarimetric survey of 22 small main-belt asteroids belonging to the Koronis ($\sim 10^9$ yrs old) and Karin ($\sim 6 \times 10^6$ yrs old) dynamical families was performed by Cellino et al. (2010). The chief aim was to compare the optical properties of asteroid surfaces of nearly identical composition but certainly modified by space weathering mechanism due to different surface exposure times. However no differences were found in the polarimetric properties of the asteroids studied. Particularly, geometric albedo derived from the polarimetric data for Koronis and Karin members turned out to be the same in the range of observational errors and closely approximated to the average albedo of S-type asteroids.

In the meantime, Rivkin et al. (2011), Thomas et al. (2011, 2012) have discovered Q-types (i.e., asteroids with OC-like spectra) among small objects belonging to the old Koronis family. The authors suggest that the observed increase in spectral slope from the smallest Q asteroids (< 4 km in diameter) to larger S asteroids originates due to space weathering effect since the small bodies may be younger than the larger. The authors themselves recognize that such an interpretation may be valid if all of the asteroids studied in their work have identical mineral composition and grain-size distribution. These constraints have not yet been verified by observations. The traits of optical maturation on asteroids with diameter from ~

0.5 to 500 km have been found from multi-color images and spatially resolved spectrometry obtained by spacecraft (Chapman, 2004; Clark et al., 2003; Ishiguro et al., 2007; Pieters et al., 2012). The maturity correlates with surface morphology (e.g., Sasaki et al., 2007; Ishiguro et al., 2007), the small-scale mixing of diverse surface components (Pieters et al., 2012), and grain size sorting (Shestopalov, 2002) rather than asteroid diameter or bulk composition.

What are the processes that "blur over" the optical manifestation of space weathering on asteroids? The most popular current ideas are: (i) exposure of fresh subsurface material by the effects of tidal forces during close encounters of asteroids with terrestrial planets (Marchi et al., 2006a; Binzel et al., 2010a, 2010b); and (ii) collisional evolution of asteroids, which steadily refreshes their surface (Richardson et al., 2005; Shestopalov and Golubeva, 2008; Paolicchi et al., 2009; Mothé-Diniz et al., 2010; Golubeva and Shestopalov, 2011; Marchi et al., 2012; Pieters et al., 2012). The first hypothesis is still being debated in the literature and refined (Nesvorný et al., 2010); the latter, we believe, is quite reasonable. The models of asteroid regoliths developed in the 1970s differ in details, but all predict high mobility of regolith layers on small bodies (e.g., Housen et al., 1979; Housen and Wilkening, 1982 and references therein). This occurs for two reasons: the low gravity of asteroids and the high frequency of inter-asteroid collisions that do not lead to asteroid disruptions but effectively rejuvenate their surfaces at the regional and global scales. Such a peculiarity of asteroid surface morphology is supported by observation from spacecraft (e.g., Sullivan et al., 2002; Richardson et al., 2005; Michel et al., 2009; Reddy et al., 2012 and others). In the case of Vesta, images from Dawn reveal that large and small impacts support rejuvenation of asteroid surface and retain original color and albedo variation without intensive coloration by weathering product of fine-grained reduced iron (e.g., Reddy et al., 2012; Schenk et al., 2012).

In this brief review, we have tried to maintain a balance between the various points of view on the optical maturation of asteroid regolith. The following substantial questions remain open:

> Can the maturation process obscure spectral-derivable mineralogical information on the potential parent bodies of ordinary chondrites?
> Why are asteroid surfaces optically less mature than lunar soil?

➤ Why is there severe disagreement between the timescales for spectral reddening inferred from laboratory ($<10^6$ yr) and asteroid investigations ($\sim 10^8 - 10^9$ yr)?

Below we discuss the issues and demonstrate that the inter-asteroid collisions effectively hamper the progress of optical maturation of asteroid-sized bodies. In the next Section, we consider the maturation process with time in the case of an undisturbed surface.

2. Optical maturity degree of asteroids and the Moon

In order to simulate the optical maturation effect on initially fresh material we use the geometrical-optics model developed by Shkuratov et al. (1999). In this model, the albedo of a particulate surface, $A(\lambda)$, is algebraic function of optical density, $\tau(\lambda)$, of some "average" particle of light-scattering material, $A(\lambda)=F_1(e^{-\tau(\lambda)})$. Notably that $\tau(\lambda)=F_2[A(\lambda)]$, i.e., the analytical reciprocity of the model permits estimating the wavelength behavior of optical density from spectral albedo. The functions $F_1(\cdot)$ and $F_2(\cdot)$ can be taken from the cited references. If the grains of reduced iron (Fe^0) are embedded into the particle rims and the iron grain size is much less than wavelength of incident light, then (Shkuratov et al., 1999; Hapke 2001):

$$\tau(\lambda) = \tau_h(\lambda) + \tau_{Fe}(\lambda) = \alpha_h(\lambda)l + \beta_{Fe}(\lambda)b, \qquad (1)$$

where $\tau_h(\lambda)$ and $\tau_{Fe}(\lambda)$ are the optical densities of a host material and reduced iron, $\alpha_h(\lambda)$ and $\beta_{Fe}(\lambda)$ being specific absorption coefficients of the host material and Fe^0, respectively; and l is the mean photon path length through the particles. In turn, $b = 2\delta c_1$, where δ is the thickness of the particle rim and c_1 is the volume concentration of Fe^0 in this layer. We used Eqs. (11, 12) from Hapke (2001) and optical constants of iron by Johnson and Cristy (1974) to calculate $\beta_{Fe}(\lambda)$; the dimension of the quantity is μm^{-1}.

Because c_1, δ, and l parameters are certainly unknown for the particles on planetary surfaces it makes sense to work with the b parameter. As follows from Eq. (1) the b parameter reduced to the length of 1 μm (i.e., $b/1$ μm) is a measure (in the sense of SMFe abundance) of the maturation level of a surface.

Optical effect of space weathering of airless planetary surfaces can be presented on the albedo $A(750$ nm$)$ – color $C(950/750$ nm$)$ plots, which have been successfully utilized in the investigation of maturation processes on lunar areas, lunar nearside (Lucey et al., 1998; Starukhina and Shkuratov, 2001), and asteroid Eros (Murchie et al., 2002; Shestopalov, 2002). A similar diagram for the Moon, S asteroids, and stony meteorites has been constructed by Golubeva and Shestopalov (2003).

Here we represent an update version of this diagram (Fig. 1) in accordance

10

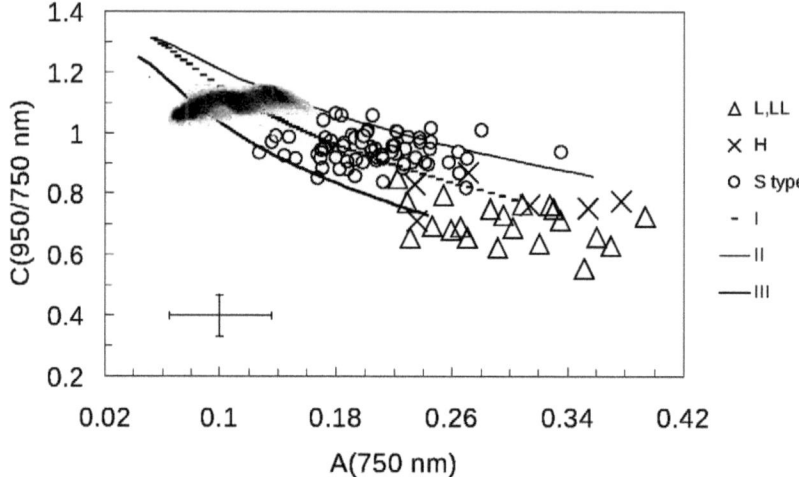

Fig. 1. Albedo-color diagram for the lunar nearside (a feature as a black birdie), S-type asteroids, and L, LL, and H ordinary chondrites (OC). Maturation trends (I, II, III) in the albedo and colors arise because of increasing SMFe concentration, c_i, within initially unweathered particles without changing their sizes (II) and with moderate decreasing particle sizes (I and III). The maturity degree, $b \propto c_i$, varies from 0 (OC domain) to 0.04, 0.05, and 0.06 for trajectories I, II, and III, respectively. The typical errors in the albedo and color estimation are also shown. Initial data for the Moon, meteorites, and S-type asteroids were taken from Starukhina and Shkuratov (2001), Gaffey (1976), and PDS SBN database (http://pdssbn. astro.umd. edu/), respectively.

with new spectral data on S asteroids available at NASA PDS Archive (http://pdssbn.astro.umd.edu/). The albedo and color of ordinary chondrite specimens without the feature of terrestrial weathering were derived from their reflectance spectra obtained by Gaffey (1976). The distribution of C(750/950 nm) against A(750 nm) for lunar nearside was taken from Starukhina and Shkuratov (2001). Albedo of S asteroids and meteorite specimens at λ = 750 nm were recalculated in the lunar albedo scale by the method described by Golubeva and Shestopalov (2003). Errors in position of asteroids and meteorites on the albedo – color plot arise due to the combination of measurement accuracy and the error of albedo scale calibration.

The lunar, asteroid's and meteorite's domains on the A(750 nm) - C(950/750 nm) plot (Fig. 1) represent material with different degree of the space weathering. Maturation trajectories, shown in this Figure, arise due to the fact that the b parameter increases from 0 (fresh OC material) through the S-asteroid region with the intermediate maturity degree, $b_A \sim 0.002 - 0.013$, to $b_M \sim 0.02 - 0.034$ within lunar region with high maturity degree. Errors in albedo and color give uncertainty in the surface maturity degree estimates ~ 0.003. This is acceptable level of accuracy to

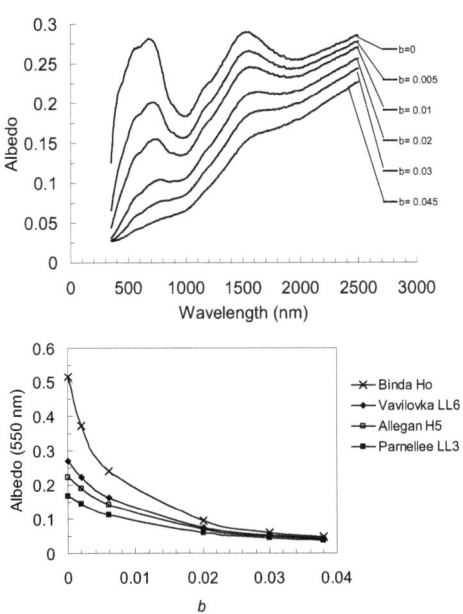

conclude that the maturity degree of the Moon as a planet is higher than that in the asteroid belt. Maturation process is capable to lead albedo A(750 nm) and color C(950/750 nm) of the ordinary chondrites to the values typical of S asteroids. Despite the fact that the composition of lunar surface is strongly different from that of ordinary chondrites, maturation process eliminates this distinction in terms of the albedo and color. In this context, the upper limit of maturity degree for the S-asteroid domain could be less than b_{Amax} ~ 0.013 because of possible difference in the composition of the S asteroids and ordinary chondrites studied in the present work.

Fig. 2. (*Top panel*) Transformation of reflectance spectra of the Vavilovka LL6 ordinary chondrite meteorite with increasing the maturity degree b. (*Bottom panel*) Saturation of the visual albedo (λ = 550 nm) of stony meteorite specimens with increasing the b parameter. Initial data from Gaffey (1976) and RELAB Database (http://www. planetary. brown. edu/ relab/).

Figure 2 shows alteration of typical ordinary chondrite spectrum and visual albedo of meteorite specimens with increasing the maturity degree b. As the absorption coefficient β_{Fe} in Eq. (1) rises steeply towards short wavelengths, all attributes of the surface optical maturation process become apparent first in the visible range of spectrum. For this reason, the albedo of the samples decreases rapidly with increasing maturity degree and approaches to 0.05 irrespective of its original value. The originally "chondritic" spectrum becomes similar in overall shape and slope to that of S-type asteroids, and the visual albedo of weathered meteorite specimens becomes close to the S-type one when the maturity degree b even less than 0.006.

Maturation effect is known to reduce absorption band intensities in the spectrum of weathered surface. First of all, surface maturation tends to erase weak

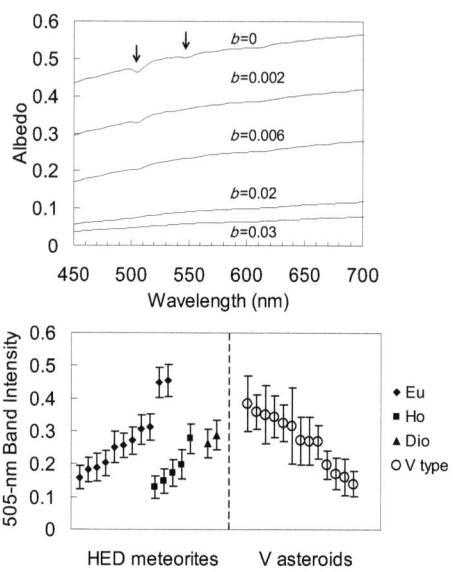

HED meteorites   V asteroids

Fig. 3. (*Top panel*) Variation of the intensity of spin-forbidden bands of Fe^{2+} cation near 505 and 550 nm in the spectrum of Binda howardite with increasing the maturity degree *b*. (*Bottom panel*) Full intensity of the 505-nm absorption band (i. e., the equivalent width in nanometers) in the spectra of HED meteorites and V-type asteroids. Original data were taken from RELAB Database and Shestopalov et al. (2007, 2008).

spin-forbidden absorptions located in visible region spectra (Fig. 3, *top panel*). But the intensity of a weak band near 505 nm in the spectra of fresh surfaces of HED meteorites turns out to be practically the same as in the spectra of V asteroids with surface compositions similar to HED (Fig.3, *bottom panel*). This absorption band, attributable to Fe^{2+} cations in pyroxenes, has been found in the spectra of many asteroids rich in mafic minerals and belonging to various optical types (Cloutis et al., 2010). Thereby these observational facts also testify favorably to the low maturity degree of asteroid surfaces.

Neither albedos or color indexes nor spectral slope or band intensities are the spectral diagnostic characteristics of asteroid compositions. Because of the specific spectral properties of the mafic mineral assemblages of ordinary chondrites (whose mafic minerals occupy a restricted range of compositions and abundances), the ordinary chondrites occupy a compact polygonal region on the plot of 950-nm band position – band area ratio (Fig. 1 in Gaffey et al., 1993). Such a correspondence between band position and BAR for these meteorites is a spectral diagnostic criterion for searching for bodies with ordinary chondrite compositions among S asteroids. We simulated the variation of these spectral characteristics of the ordinary chondrite specimens depending on surface maturity degree and present the results of the simulation in Fig. 4. When maturity degree *b* changes in the range of 0 – 0.02, the variations of the spectral characteristics for the maturation effect remain within the OC region on the

13

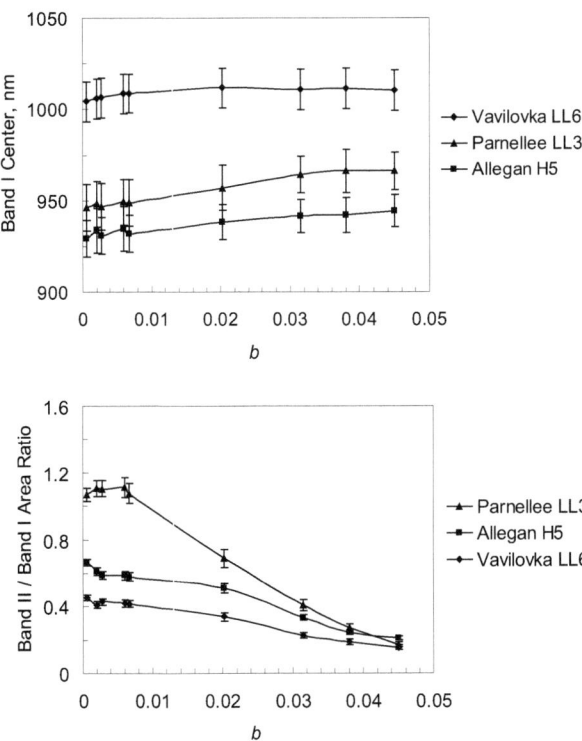

Fig. 4. Variations of 950-nm band center position and band area ratios for absorption bands centered near 950 and 1900 nm in the spectra of ordinary chondrites against the surface maturity degree b. The original meteorite spectra were taken from Gaffey (1976).

above mentioned plot (Gaffey et al., 1993). Thus, the space weathering intensity in the main belt asteroids ($b_A \ll 0.02$) does not impede spectral detection of the ordinary chondrite assemblages on S-asteroid surfaces using band position and band area ratio. Recently, Gaffey (2010) came to the analogous conclusion based on the results of laboratory simulation of maturation effect on the spectra of mafic silicate targets.

3. Timescales of optical maturation of undisturbed surface: data of laboratory experiments

The exposure time necessary to change the optical properties of lunar or asteroidal surfaces by space weathering have repeatedly been derived from laboratory experiments. However, these estimations are very variable. For example, Hapke (2001) supposes the time required to convert initially fresh chondritic surface into S-asteroid regolith under the action of solar wind sputtering is not less than 50000 years. Loeffler et al. (2009) infer the weathering time of about 5000 years at 1 AU from He ion irradiation of olivine powder sample. Sasaki et al. (2001), using nano-pulse laser shots at olivine pellet samples, estimated a weathering time of ~ 10^8 years at 1 AU and ~ 6×10^8 years in the asteroid belt; this result was corroborated by Brunetto et al. (2006b). The weathering timescale of asteroid surfaces is supposed to be of the order of, or less than, 10^6 years (Brunetto et al., 2006a; Vernazza et al., 2009).

3.1. Solar wind as a source of surface maturation

In his experiment, Hapke (1973) exposed silicate rock powders to proton irradiation in a high vacuum environment to simulate darkening of lunar surface materials by solar wind. The total irradiation of the samples was measured as the electric charge density (Coulomb cm^{-2}). Since the average solar flux is about 3×10^8 protons cm^{-2} s^{-1} at 1 AU, Hapke (1973) derived a simple ratio between the charge density and the exposure time by solar proton irradiation: ~5 C cm^{-2} or 10^4 yr, ~30 C cm^{-2} or 6×10^4 yr, and so on. We used this ratio to modify Figure 3 from the work by Hapke (1973) and plotted a graph of visual albedo of the samples versus the exposure time t.

Based on the works of Morris (1977; 1980) we can suppose that the volume fraction of the fine-grained metallic iron is a linear function of time, i.e., $c_1(t) = c_{1 \, max}(t/T)$, where $c_{1 \, max}$ is maximal concentration of Fe^0 accumulated for the time T, when visual albedo of the surface saturates. So we can write the surface maturity degree as follows:

$$b(t) = 2\delta c_{1max}\left(\frac{t}{T}\right) = b_{max}\left(\frac{t}{T}\right). \tag{2}$$

Then we calculated theoretical relations between albedo and t using Eq. (2) and Shkuratov's model.

As is seen from Figure 5a, albedo of the fine-grained samples varies faster with exposure time than that of the coarse-grained ones. Apparently, this occurs due to a more porous surface of the fine-grained samples, so that vapor-deposited coatings on the particle surfaces form faster. After time $T \sim 2 \times 10^5$ yr, visual albedo of the olivine basalt powders subjected to proton irradiation changes very slowly. Figure 5b illustrates the fact that, all other things being equal, regardless of the initial albedo of unaltered samples, this value approaches ~ 0.05 for about the same time interval T, which can be called saturation time of solar wind darkening at 1 AU. We corroborate the finding of Hapke (1973) that the undisturbed lunar regolith would darken in $\sim 10^5$ yr. This result is also supported by numerical simulation of the sputtering of the lunar soil by solar wind protons (Starukhina, 2003) and Monte Carlo computation of complex dynamical evolution of an uppermost layer of the lunar regolith, the particles of which should show the solar wind saturation effect (Borg et al., 1976).

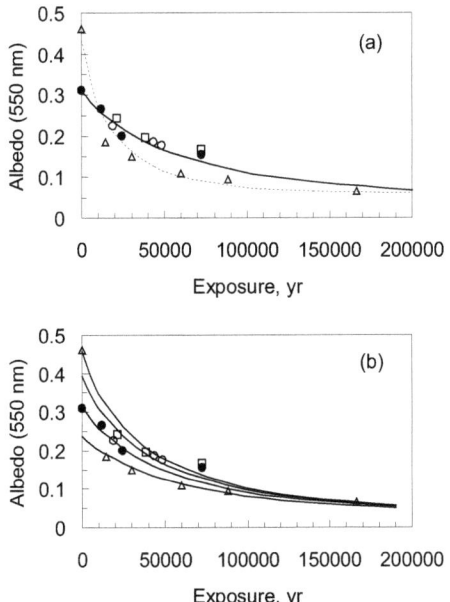

Fig. 5. Albedo of olivine basalt powder versus the exposure time for varying H-ion beam densities and particle sizes in simulations of the darkening of silicate powders by solar wind irradiation at a heliocentric distance equal to 1 AU. These original data were adopted from Hapke (1973): filled circles denote ion beam density equal to 0.33 mA cm^{-2} and particle size less than 37µm; open circles: 0.17 mA cm^{-2}, < 37µm; squares: 0.09 mA cm^{-2}, < 37µm; triangles: 0.50 mA cm^{-2}, < 7µm; (a) Dashed line fits the fine-grained samples at b_{max} = 0.08 µm and $T=2 \times 10^5$ yr whereas solid line fits the coarse-grained samples at b_{max} = 0.032 µm and $T=2 \times 10^5$ yr; (b) Numerical calculations for various albedos of host material as a function of the exposure time were carried out at b_{max} = 0.038 µm and $T=2 \times 10^5$ yr.

As is seen from Figure 2, the originally "chondritic" spectrum became similar in overall shape and slope to that of S-type asteroids at $b \approx 0.002 - 0.006$ or, equivalently, between 1×10^4 and 3×10^4 yr on the above timescale. For the asteroid belt, we can recalculate this spatter – darkening time using the relation derived by Hapke (2001):

$$t_{MB} = t_{1AU} a^2 \sqrt{1-e^2} \, ,$$

where a and e are the semimajor axis in AU and the eccentricity of an asteroid orbit. For an asteroid in the middle of the main belt with $e = 0.14$, t_{MB} ranges from 7×10^4 to 2.2×10^5 yr or $\sim 1.5 \times 10^5$ yr on average.

3.2. Microparticle bombardment as a source of surface maturation

Laser irradiation experiments that simulate micrometeorite impact vaporization of regolith particles show that the timescale of optical maturation by this process seems to be longer than the time of solar wind darkening (Sasaki et al., 2001; Brunetto et al., 2006b). To convert the spectrum of an originally fresh olivine target bombarded by micrometeorites with a diameter of 1 μm and an impact velocity of 20 km/s into an A-type asteroid spectrum requires $\sim 10^8$ yr in space at 1 AU distance (Sasaki

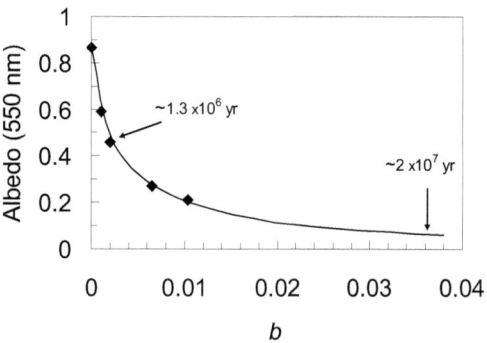

Fig. 6. Albedo of olivine pellet samples before and after pulse laser irradiation in a simulation of space weathering effect by micro-particle impacts (Sasaki et al., 2001) versus parameter b which has been estimated by Shestopalov and Sasaki (2003). The five albedos (diamonds) correspond to (from top to bottom): non-irradiated sample, 15-mJ irradiation, 30-mJ irradiation, five and ten repetitions of 30-mJ irradiation treatment. Numerical calculation (line, this work) was carried out at $b_{max} = 0.037$ μm and $T=1.9 \cdot 10^7$ years.

et al., 2001). We recalculated this time interval taking into account the cumulative flux of dust particles in the range of masses from 2×10^{-15} to 10^{-7} g (approximately 0.12 – 46-μm size particles) at 1 AU taken from McBride and Hamilton (2000). The same spectral alteration (i.e. the spectrum of olivine $\rightarrow$ roughly A-type spectrum) takes $\sim 1.3 \cdot 10^6$ yr assuming the impact velocity to be 20 km s^{-1}. The reduction of the time interval occurs mostly due to rarer but heavier dust particles of ~ 45 μm in size.

Figure 6 shows our interpretation of the laser irradiation experiment of Sasaki et al., (2001). The 30-mJ irradiation of the olivine pellet at $b \approx 0.002$ corresponds to 1.3×10^6 yr in space at 1 AU; the saturation time for this process is $\sim 2 \times 10^7$ yr, which is greater by two orders of magnitude than that for solar wind darkening at the same distance from the Sun. Visual albedo of the altered olivine samples at 1.3×10^6 yr is still far from the saturation level of ~ 0.05.

It is difficult to recalculate correctly the timescale for microparticle impact darkening in the case of the main belt asteroids since neither a size frequency distribution for microparticles nor their impact velocity are well known there. For example, to evaporate a unit mass of forsterite, an impact velocity of 10 km/s or higher is required (Hapke, 2001 and reference therein). The average relative velocity of asteroids is only ~ 5 km/s and only about 3% of all impacts in the main belt occur at velocities over 10 km/s (O'Brien et al., 2011). The sources of the high-velocity dust particles in the main belt are also unknown today. It appears that the role of microparticle impacts into main belt asteroids is reduced to the crushing of small regolith particles and, possibly, to their partial vitrification.

3.3. Summary

Thus, we conclude that solar wind is apparently the sole and constantly operative process, which could lead to optical maturation of asteroid surfaces. This short weathering timescale occurring due to the solar wind irradiation (our estimate $\sim 1.5 \times 10^5$ yr or $< 10^6$ yr according to other investigators (e.g., Vernazza et al., 2009) leads to the following contradiction. Indeed, the undisturbed surface of a body having ordinary chondrite composition will acquire the spectral features of an S-type asteroid within $10^5 - 10^6$ yr. Since the surface maturation of such an asteroid will extend beyond this time, its visual albedo will be reduced to ~ 0.05 by the saturation time of $\sim 1.4 \times 10^6$ yr and its reflectance spectrum will become reddish, akin to the lowermost spectral curve in Figure 2. Both these time intervals are much shorter than a collisional lifetime of asteroids larger than 8 km in diameter ($\sim 5 \times 10^9$ yr, O'Brien and Greenberg, 2005). Consequently, instead of the S-type population we should observe numerous low-albedo asteroids having reddish spectra with subdued absorption bands. This does not happen, of course. The key words for solving this logical contradiction are "undisturbed surface". Such an approach is reasonably sufficient for the Moon but not for asteroids.

As is evident from simulation of the exposure history of lunar regolith particles (Borg et al., 1976 and references therein), the lifetime of a freshly deposited layer before burial by another layer is ~2.5×10^7 yr. As a result of random walks under meteorite impacts, some particles sometimes reach the uppermost layer of the surface, while others descend. The average exposure time of any surface particle to the solar wind is ~ 5000 yr; and the number of exposures of a particle to the solar wind is ~ 20 – 30, so the total exposure time of mature particles is ~ $10^5 - 1.5 \times 10^5$ yr. The thickness of the uppermost layer of the lunar regolith, consisting of the optically matured particles, is ~ 5 mm. This "skin" (the term by Borg et al., 1976) almost everywhere covers the lunar surface. Largely due to this circumstance, the Moon has low albedo and reddish spectrum. Thus the concept of "undisturbed surface" is quite applicable to the Moon because of the low frequency of meteorite impacts and the short timescale of solar wind darkening. A different situation occurs in the case of asteroids. The timescale necessary to convert a "chondritic" spectrum into S-type is approximately the same as for lunar mature regolith (i.e., ~10^5 yr). However, one can expect the low gravity of asteroids and the high frequency of non-catastrophic collisional events result in an essential difference in intrinsic mobility of asteroid and lunar regoliths and, as consequence, in difference in their optical maturity degree.

In the next Section, we examine the main traits of the evolution of asteroid regoliths within the framework of the problem. It should be stressed that our intent here is not a detailed simulation of the asteroid regolith formation but to illustrate the fact that the continuous collisional resurfacing of asteroids hampers strong optical alteration of their surfaces by solar wind.

4. Collisions between asteroids

The spectrum of an initially fresh surface exposed to solar wind and having ordinary chondrite composition will exhibit some resemblance to the spectra of S-type asteroids in the time of $T_{OC \to S} \approx 1.5 \times 10^5$ yr at a distance of about 2.6 AU. At the same time, the surface of the asteroid undergoes numerous collisions with bodies of different masses and sizes. What happens to asteroid surfaces during this relatively short time span?

4.1. The number and frequency of collisions

The number of collisions occurring between a target asteroid with diameter D_a and projectiles with diameters equaled to or larger than D_p during a time interval τ can be estimated by the following equation (e.g., Wetherill, 1967; Farinella et al., 1982):

$$N_p = 0.25 P_i D_a^2 N(> D_p) \tau, \tag{3}$$

where P_i is the intrinsic collision probability, $N(> D_p)$ is the cumulative number of bodies larger than D_p in the projectile population. All bodies are assumed to be spherical and their diameters are measured in km. If N_p equals 1 in Eq. (3) then τ is the mean time between impacts; the mean frequency of the impacts is simply expressed in terms of $T_{OC \to S}$, the time span of interest:

$$\frac{1}{\tau_p} = \frac{N_p}{T_{OC \to S}}. \tag{4}$$

In Eq. (3), P_i may be interpreted as the collisional rate between two asteroids for which the sum of their radii is 1 km. We are interested in non-catastrophic collisions of asteroids, so the diameter of target asteroids is much larger than that of projectiles and a component including D_p^2 is omitted in Eq. (3). Calculation of P_i is a non-trivial task for the elliptic and noncoplanar orbits of asteroids. Beginning with the pioneering works of Öpik (1951) and Wetherill, (1967) the intrinsic collision probability for various populations of asteroids both in the main belt and beyond its bounds has been estimated in many studies (e.g., Davis et al., 2003). In our work the value for the main belt asteroids was taken from Bottke and Greenberg (1993), $P_i = 2.86 \times 10^{-18}$ km^{-2} yr^{-1}.

To calculate $N(>D_p)$ we used an incremental size distribution for projectile population (O'Brien and Greenberg, 2005) derived from modeling collisional and dynamic evolution of the main-belt and near-Earth asteroids. This model size distribution includes asteroids up to 0.001 km in size and was tested against a number of observational constraints such as the observed population of the main-belt asteroids and NEAs from various surveys, the cosmic ray exposure ages of meteorites, and the cratering records on asteroids.

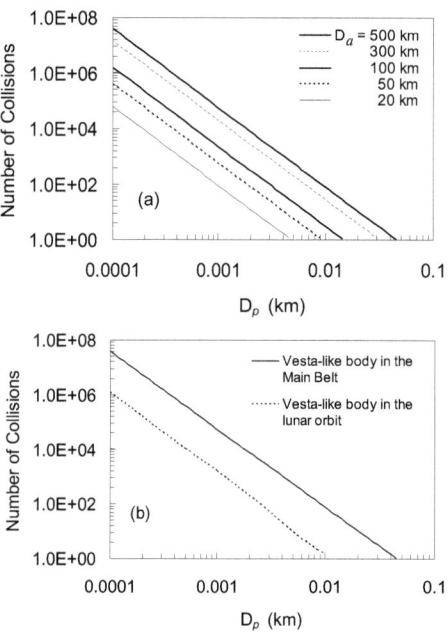

Fig. 7. (a) The number of collisions between asteroids of given diameter (D_a) and impactors of diameter (D_p) or larger in time interval $T_{OC \to S} = 1.5 \times 10^5$ years; (b) The number of collisions between 500-km body and impactors in the asteroid belt and in the lunar orbit during the same time interval. The dotted line is linear extrapolation based on the NEA size-frequency distribution taken from Werner et al. (2001);

Figure 7a illustrates the number of asteroid collisions in the time $T_{OC \to S}$ depending on the diameter of projectiles. This is, in fact, the asteroid size frequency distribution (O'Brien and Greenberg, 2005) inverted to N_p using Eq. (3). The model does not indicate a deficit of small asteroids in the main belt therefore we expanded our calculation up to impactor diameters of 0.0001 km. The diameter of the largest non-catastrophic impactor is defined by both the time period (i.e., $T_{OC \to S}$ in the given case) and the asteroid cross section, and varies from 0.0045 to 0.045 km for target asteroids from 20 to 500 km in size. As is seen from Figure 7a, even in this relatively short time span, asteroids undergo a great variety of impacts, which, of course, do not lead to total disruption of asteroid bodies, but effectively garden their surfaces.

Figure 7b demonstrates an essential difference in a "bombardment hazard" in the asteroid belt and in the lunar orbit. For that, we used an average lunar impact probability of 1.86×10^{-10} yr^{-1} per asteroid as derived by Werner et al. (2002) for all known near-Earth asteroids with absolute magnitude $H \leq 18^m$. To calculate the number of collisions between a 500-km body and projectiles in the lunar orbit we must take into account, firstly, the difference in the lunar and body cross section (i.e., factor of D_B^2 / D_{MOON}^2) and secondly, the fact that the cumulative size-frequency distribution of projectile population near 1 AU (Werner et al., 2002) was scaled to $N(>D_p) = 1$ at $D_p = 1$ km. Fortunately, since $N(> 1~km)$ is known and equals ~ 700, we have finally:

$$N_p\big|_{at1AU} = 1.86 \times 10^{-10} \times 700 \times \frac{D_B^2}{D_{MOON}^2} \times N(> D_p) \times T_{OC \to S},$$

where $N(>D_p)$ is the cumulative number of projectiles near Earth. In the model used, the smallest projectile diameter of the near-Earth population is only 10^{-2} km thus we were forced to extrapolate our calculations in the range of projectile diameters typical of the asteroid belt. The dotted line in Figure 7b is the result of such an extrapolation. We cannot state that the forecast outcome for a Vesta-size body in a lunar orbit is sufficiently precise. However, it is clear that the bombardment hazard is much less in the lunar orbit than that in the asteroid belt. This is the most essential condition for the retention of a thin weathered layer on the lunar surface. As a result, the average visual albedo of lunar nearside is low (~ 0.1) and its average reflectance spectrum is extremely red.

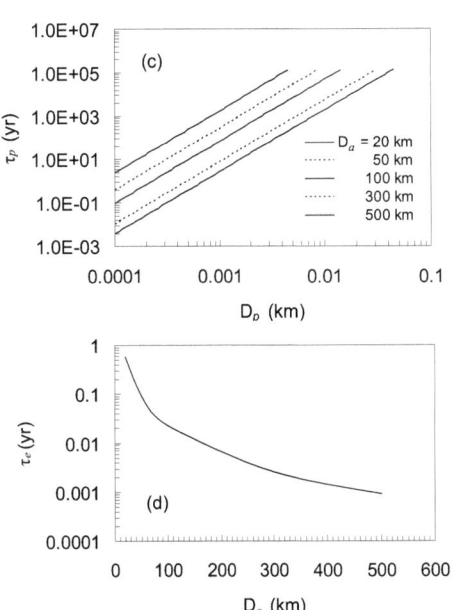

Fig 7. (c) The mean time interval between impacts versus projectiles of the given size or greater; (d) The expected time between impacts for asteroids of various diameter.

The time interval τ_p between impacts as a function of the projectile diameter for the asteroids of various sizes is plotted in Figure 7c. As is seen from this Figure, the larger the target asteroid the shorter τ_p; the smaller the size of the impactors the higher the frequency of collisions. We can estimate an expected time of collision, τ_e, of an asteroid with *any body* of the target population. For the given target asteroid, it is simply $T_{OC \to S}$ divided by the total number of impacts, $N_{coll} = \sum N_p$, in this time period. In accordance with Eq. (4)

$$\tau_e = \frac{1}{\sum \left(\dfrac{1}{\tau_p} \right)} ,$$

(5)

where the index p denotes the projectile population with diameter of D_p. The τ_e dependence on asteroid diameter is shown in Figure 7d. A body of 20 km in size may undergo one impact in the expected time of about 210 days; the 50-, 100-, 300-, and 500-km asteroids do in ~ 33 days, 8 days, 22, and 8 hours, respectively. Therefore, asteroid surfaces cannot be considered as undisturbed, they rather are "impact-activated".

A heavy shower of impactors generates craters of various sizes on asteroids and triggers impact-induced seismic processes.

4.2. Craters and crater ejecta

We will assume for the moment that all impacts into asteroids produce craters. In reality this may not be the case, because some projectiles can move along trajectories nearly tangential to the impact site and perhaps create grooves, but not craters in the usual sense of the term. However, even if the number of such very oblique impacts will be one-half of the total impacts, it has little effect on the order of magnitudes which we utilize in this work.

One of the approaches to solving the tasks of the impact dynamics is scaling laws. This phenomenological approach requires some explanation (Housen et al., 1983; Holsapple, 1993 and reference therein). The scaling laws determine the functional relations between the outcomes of hypervelocity impacts (say, radius and depth of craters, the range of crater ejecta and thickness of ejecta deposits) and input parameters such as impact velocity, characteristics of impactors, material strength, gravity, and others. Experimental studies of impacts, analytical solutions and code calculations of the relationships between mass, momentum, and energy, as well as the dimensional analysis of the variables, are the primary tools to derive scaling theories. In such a way, the scaling laws provide a bridge between laboratory experiments and large planetary impact events, which are impossible to reproduce in actual practice.

We consider an impact occurring in a gravity field of magnitude g. When the strength of the target material predominates over gravitational effects, the latter plays a negligible role in the process of crater formation. Otherwise, the gravitational forces regulate the cratering process. These two extremes are often referred as "strength-dominated and gravity-dominated regimes" depending on the scale of the impact event. In our case, the diameters of most projectiles vary from ten centimeters to some meters, so that craters on asteroids will form in the strength regime at small sizes and tend to transition to the gravity regime at larger crater sizes. We use simple scaling relationship between the crater and projectile diameter taken from Richardson et al. (2005):

$$D_{cr} = 30D_p. \tag{6}$$

Judging from Figure 16 in the referred work, this relation corresponds to transient conditions between pure strength and pure gravity regimes for various types of target materials from loose sand to competent rocks and well satisfies the numerical hydrocode simulation of the cratering events on asteroids.

Now using Eq. (6) we can estimate for the given asteroid the total area of craters S_{tcr} formed in the time $T_{OC \rightarrow S}$:

$$S_{tcr} = 225\pi \sum D_p^2 N_p ,$$

where N_p is the number of impactors with diameter D_p. As N_p varies directly with D_a^2, it follows that S_{tcr} and the total area of the asteroid surface, S_a, are directly proportional. For the target asteroids of 20–500 km in size, the ratio of S_{tcr}/S_a is approximately the same and equals ~0.004. We verified this inference using other power-law scaling approximations that link crater diameters, impact velocity, and gravitational acceleration (Eq. 16 in Housen et al., 1979; Eq. 22a in Holsapple, 1993) and obtained similar results: the craters forming in $T_{OC \rightarrow S}$ ~1.5×10^5 yr occupy less than 1% of the asteroid surface even if they do not overlap. That is, the impact flux has not yet had time to destroy the original surface of asteroids completely. Constancy of S_{tcr}/S_a ratio for asteroids of various sizes reflects the fact that the small-crater population dominates among the craters in question.

A portion of the kinetic energy of the impactor (about 10 %) is spent on ejection of material during crater formation (O'Keefe and Ahrens, 1977). Dimensional analysis leads to the following dependence of the velocity, v, of excavated material when it passes through the original target surface at a distance, x, from the impact point (Housen et al., 1983; Housen and Holsapple, 2011):

$$\frac{v}{U} = \left[\frac{x}{R_p} \left(\frac{\rho}{\delta} \right)^v \right]^{-1/\mu} f(x/R_{cr}), \tag{7}$$

where U is the impact velocity, R_p is the impactor radius, ρ and δ are the mass densities of the target and impactor materials, respectively, R_{cr} is the crater radius, and $f(\cdot)$ is some function that is not determined by means of dimensional analysis. Housen and Holsapple (2011) stress that Eq. (7) describes the ejection velocity as a function of a launch position x both for the strength and for the gravity regimes, as the crater radius R_{cr} depends on the strength and gravity effects. The authors note also that the power-law approximation with exponent $-1/\mu$ does not work near the crater edge where the ejection velocity goes to zero. It is therefore assumed that the function $f(\cdot)$ in Eq. (7) can be expressed as $(1 - x/n_2 R_{cr})^p$. That is, Eq. (7) takes the form (Housen and Holsapple, 2011):

$$\frac{v}{U} = C_1 \left[\frac{x}{R_p} \left(\frac{\rho}{\delta} \right)^v \right]^{-1/\mu} \left(1 - \frac{x}{n_2 R_{cr}} \right)^p, \tag{8}$$

where constant C_1 and exponents v, μ, p should be found from impact experiments; n_2 takes separate values $n_{2,S}$ and $n_{2,G}$ for the strength and gravity regimes; x varies from $n_1 R_p$ to $n_2 R_{cr}$ and n_1 is assumed to be approximately 1.2.

The scaling law for the mass of ejected material having launch position less than x is $M(<x) = k\rho x^3$ (Housen and Holsapple, 2011). But $M(x)$ goes to zero if $x \rightarrow n_1 R_p$ because material moves in a downward direction inside of $x = n_1 R_p$ and is not ejected. With this in mind the authors modify this expression to the form:

$$M(<x) = k\rho \left(x^3 - [n_1 R_p]^3 \right), \tag{9}$$

where k is a constant that should be estimated from experiments. The authors point out that $M(<x) = M(>v)$, that is, the ejected material has velocity greater than arbitrary value, v, because the ejection velocity decreases monotonically with increasing x according to Eq. (8).

In this scaling theory, the total mass of ejecta and the crater radius are coupled variables:

$$M_{cr} = k_{cr} \rho R_{cr}^3, \tag{10}$$

where, k_{cr} is an empirical constant less than 1 (see the referred work for more details). The scaled mass of ejecta which is launched faster than the corresponding velocity is

$$\frac{M(>v)}{M_{cr}} = \frac{k}{k_{cr}} \left(\frac{x^3}{R_{cr}^3} - \frac{[n_1 R_p]^3}{R_{cr}^3} \right). \tag{11}$$

Owing to the simple relation between the crater and projectile diameter (Eq. 6), which we use to characterize the impact events on asteroids, the ejection velocity and the scaled mass of ejecta depend neither on the size of crater nor the impactor size. Since $R_p = \frac{1}{30} R_{cr}$ and the launch position x is measured in units of crater radius (i.e., $x = qR_{cr}$), then Eqs. (8) and (11) are reduced to

$$\frac{v}{U} = C_1 \left[30q \left(\frac{\rho}{\delta} \right)^v \right]^{-1/\mu} \left(1 - \frac{q}{n_2} \right)^p, \tag{12}$$

$$\frac{M(>v)}{M_{cr}} = \frac{k}{k_{cr}} \left(q^3 - [n_1/30]^3 \right), \qquad n_1/30 \le q \le n_2 . \tag{13}$$

Distribution of the scaled mass of material which is ejected faster than the given velocity v is shown in Figure 8. The calculations were performed with the following assumptions. For the sake of simplicity we suppose $\rho = \delta$. The impact velocity U is equal to 5 km/s, average relative velocity in the asteroid belt. The numerical values of the empirical parameters were chosen according to the data listed in Table 1 from Housen and Holsapple (2011): $C_1 = 0.5$, $\mu = 0.4$ and $p = 0.3$ – they are all for porous, regolith-like materials; $k/k_{cr} = 0.5$ is reasonable for common soils and could be the upper limit for asteroid porous regolith; $n_1 = 1.2$ and $n_2 = 1$.

Figure 8 illustrates the important property of the ejecta model developed by Housen and Holsapple (2011): at least for small impact events when the simple "cube-root" scaling law (Eq. 6) works, almost all the mass of the impact crater ejecta accumulates on the target asteroids with diameter larger than 10 km. Only a small

26

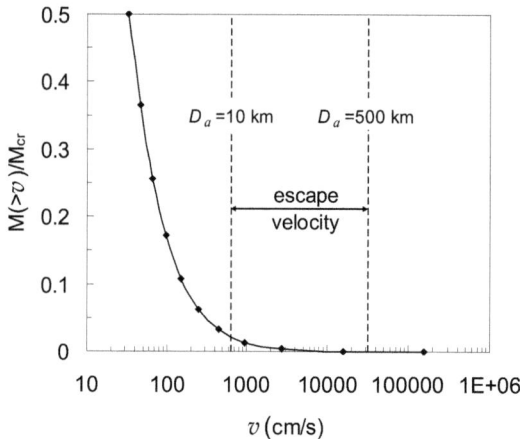

Fig. 8. Scaled mass of ejecta with velocity greater than v against the ejection velocity. The range of an escape velocity for asteroids with diameter between 10 and 500 km is also shown.

fraction of material launched from near the crater center moves faster than the escape velocity v_{es}, and can leave the asteroids. The substantial regolith layer seen covering small asteroids such as Gaspra, Eros, and Ida is a good corroboration of this conclusion (Sullivan et al., 2002).

If the launch position varies in the range $x_1 \le x \le R_{cr}$, the volume, ΔV, of material ejected from the given crater is

$$\Delta V = \frac{1}{\rho} \Delta M = \frac{1}{\rho} \left[M(< R_{cr}) - M(< x_1) \right] = k R_{cr}^3 (1 - q_1^3), \qquad (14)$$

where q_1 corresponds to the ejecta velocity $v(q_1) \sim 0.9 v_{es}$ and $k = 0.3$ in accordance with the data from Table 1 in Housen and Holsapple (2011). The total volume of material produced by impacts into the given asteroid is

$$\Delta V_t = \sum \Delta V N_p = 15.8 (1 - q_1^3) \sum D_p^3 N_p. \qquad (15)$$

For simplicity, we assume that after a lapse of a sufficiently long time period the ejecta blankets cover asteroid surfaces more or less uniformly. Then the average thickness of this layer on the asteroid with diameter D_a is approximately

$$H_b \approx \frac{\Delta V_t}{\pi D_a^2}. \qquad (16)$$

We plotted Figures 9 and 10 using Eqs (12–16) and the same numeric values of the foregoing empirical parameters. Shown in Figure 9a is the dependence of total volume, ΔV_t, of ejecta on the time interval τ between impacts. As is seen from this Figure and Figure 7c, just small impactors provide the main growth in ΔV_t due to the

27

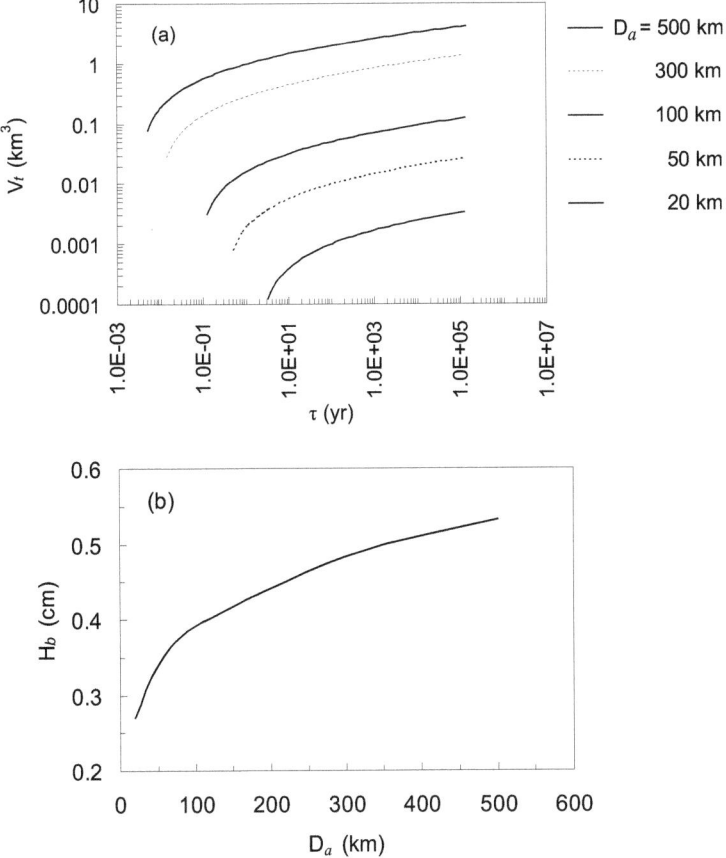

Fig. 9. (a) Volume of material (V_t) ejected during crater formation on asteroids of various diameter (D_a) as a function of the time period between impacts (τ); (b) Average thickness (H_b) of the ejecta blanket formed on asteroids of various sizes in the time of 1.5×10^5 years. The numerical values of the empirical parameters for porous, regolith-like materials were chosen according to the data listed in Table 1 from Housen and Holsapple (2011): $C_1 = 0.5$, μ = 0.4 and $p = 0.3$, $k = 0.3$, $n_1 = 1.2$, and $n_2 = 1$. The impact velocity $U = 5$ km/s, and $\rho = \delta$;

higher frequency of collisions with asteroids; the larger the cross-section of the asteroid, the larger the total volume of ejected material. The average thickness of the ejecta blanket as a function of the asteroid diameter is shown in Figure 9b. Thickness of the blankets which are accumulated on the asteroids of various sizes in the time

28

span $T_{OC \to S} \sim 1.5 \times 10^5$ yr averages from ~ 0.3 cm for the 20-km asteroid to ~ 0.5 cm for the 500-km asteroid.

To verify the correctness of our approach we compared our calculation of the thickness of asteroid regolith formed over an interval of $\sim 10^9$ yr on a 20-km body (see Figure 10) with analogous computation of Richardson (2011). If we take into

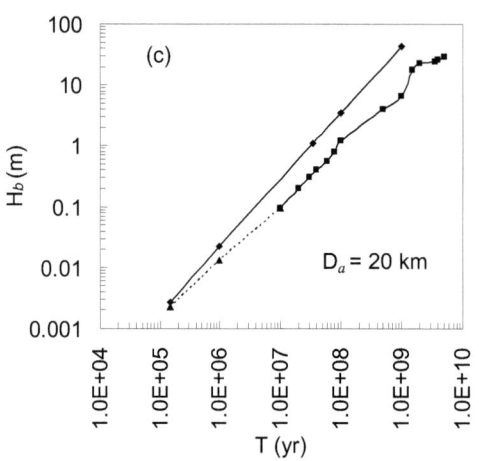

Fig. 10. The growth of average regolith depth in an Eros-sized body over the bombardment time of $\sim 10^9$ years. Diamonds, squares, and triangles denote, respectively, our calculations, the data from Richardson (2011), and extrapolation of these data towards shorter time span.

account the differences in baseline assumptions, the calculation algorithms, and the used numerical values of the empirical parameters, one can conclude that the theoretical curves shown in Figure 10 are in satisfactory agreement (i.e., both indicate accumulations of a meters to tens of meters thick regolith). Indeed, we can also deduce, as Richardson (2011) has done, that the regolith layer is generated on decimeter, meter, and decameter scales after a lapse of $\sim 10^7$, 10^8, and 10^9 yr, respectively, on asteroids of 20 km in size. Extrapolating to shorter time intervals we derive a subcentimeter scale for regolith thickness on timescale of $\sim 10^5$ yr. The result for asteroids of various sizes is represented in Figure 9b.

So, over the lifetime of asteroids, fresh subsurface material is ejected and accumulating on their surfaces due to the ongoing cratering process. On asteroids of various sizes, the thickness of this layer formed in the time of $T_{OC \to S} = 1.5 \times 10^5$ yr amounts to some thousands of micrometers. That is, the original surface, which should have been irradiated by the solar wind, is episodically covered by layers of fresh material. Penetration of visible light into silicate minerals is on the order of a few tens of micrometers, say, ~ 50 μm for wavelength of 0.5 μm. This topmost layer of asteroid regolith is formed in $\sim 5 \times 10^3$ yr on asteroids with $D_a \sim 20$ km (and even faster on larger asteroids). This time is much shorter than the $T_{OC \to S} = 1.5 \times 10^5$ yr, therefore this topmost optically active layer, as with the underlying layers, is actually immature material.

On the other hand, the layer of ejecta deposited on asteroids in the time $T_{OC \to S}$ is sufficiently thin on the body scale. Since crater formation is a stochastic process, sediment thickness can vary over asteroid surface, somewhere there may be more, somewhere less than the average. Because of natural topography, an asteroid's surface area is larger than the area of an equivalent sphere; therefore some asteroid terrains could be free from ejecta deposits. A detailed description of regolith formation on asteroids is beyond the scope of this work (see the excellent works of Richardson et al. (2005; 2007) and Richardson (2009)). Nevertheless we can infer that on the same weathering timescale, the maturation level of an asteroid as a whole will be less than that of an undisturbed and unshielded surface exposed to solar wind. Such an asteroid, in the case of disk-unresolved observations, would look like a slightly weathered body. Of course, we can try to increase the maturation level of an asteroid surface by expanding the weathering timescale, say, up to ~8×10^8 yr (this value from Paolicchi et al., 2007). However, over this time the original surface will be destroyed by ensuing impacts, the impact ejecta blanket thickness will run up to some meters, a new reference surface will be formed, and... one can start reading this Subsection from the beginning.

The formation of regolith on asteroids is accompanied not only by the ballistic motion of particles, but being already on the asteroid surface, the particles are involved in other types of movements such as downslope movement, shaking, mixing, sorting by particle size - all triggered by impact-induced seismic events.

4.3. Seismic activity of asteroids and regolith motions

A small fraction of the kinetic energy of an impactor (~ 0.1% or even less) is converted into seismic energy that in the form of seismic waves travels throughout an asteroid. Depending on projectile and asteroid masses, the impacts generate the seismic vibrations of asteroid surface in the regional or global scale. Recently, Richardson et al. (2005; 2009) have developed the theory of seismic processes following impacts into asteroid-sized bodies. Asteroids are believed to be fractured bodies having internal structure similar to upper lunar crust: a relatively thin regolith layer covers a thick megaregolith formation, which gradually turns into fractured bedrock. Each impact into an asteroid is the source of the body spherical waves, which penetrate through the asteroid interior and undergo multiple scattering in the randomly inhomogeneous medium, and refraction and reflection at the internal interfaces and asteroid surface. Owing to the small sizes of asteroids and to the fact

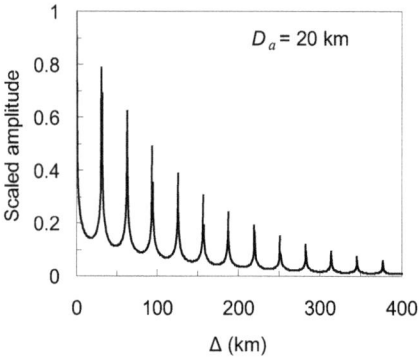

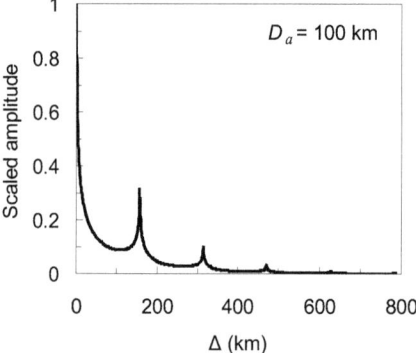

Fig. 11. The effect of multiple passages of a surface seismic wave between an impact site (the odd maxima) and its antipodal point (the even maxima) on spherical bodies with diameter of 20 and 100 km. Δ is distance between the impact site and front of the wave. Scaled amplitude of the surface wave equals 1 at $\Delta = 0$.

that the body wave velocity is about 2 – 3 km/s in fractured medium, the asteroid interior is quickly filled up with a low-frequency "seismic sound". Most asteroids are likely extremely dry as is the lunar interior, therefore the seismic energy dissipation is low and the asteroid is capable of retaining the impact excitation longer. In the case of asteroid Eros, the duration of synthetic seismograms obtained away from an impact site varies from a few minutes to about an hour depending on the impactor size (Richardson et al., 2005). Reverberations of the lunar surface that have been observed during lunar seismic experiments could last longer than an hour and half (Toksöz, 1975; Dainty et al., 1974).

When the body waves reach the surface, they generate cylindrical seismic waves that travel along an asteroid's surface within an upper shell having thickness of the order of the wavelength of surface oscillations. The surface waves are characterized, in comparison with the body waves, by a low velocity, high intensity, low-frequency spectrum, a rapid attenuation with depth, and long-term oscillations. At large distances from the impact site (or epicenter in the given context), surface waves become much intense than body waves and can cause more extensive modification of natural topography. Once originated, a surface wave diverges from the epicenter in all directions and, rounding the entire asteroid, may meet at the antipodal point where a new seismic event of lesser intensity occurs. Figure 11 illustrates the effect of the multiple passages of a surface harmonic wave between the epicenter and its antipode on spherical bodies with the same fractured structure as described above. In our illustrative example, the phase velocity of the wave and its period are 0.49 km/s and 5 s, respectively. It is supposed that the wave penetrates through a homogeneous medium, so that the wave velocity does not depend on distance from the epicenter. The maxima of the curves shown in the Figure decrease exponentially due to dissipation and scattering of seismic energy. The seismic attenuation parameter, Q = 2000, and the seismic diffusivity, K_s = 3 m^2 s^{-1}, were taken from Richardson et al. (2005). This example demonstrates that a single impact event (specially a large event) is able to induce regolith particle motions, even on the antipodal side.

Let us consider a particle motion in a surface wave of Rayleigh type. There are other types of surface waves known from terrestrial seismology, but they need more specific environments than those for Rayleigh waves to be generated (Sheriff and Geldart, 1982). If Rayleigh wave propagates through a semi-infinite elastic and isotropic layer, then each unit of the medium describes a closed elliptic trajectory in a plane normal to the surface and parallel to the traveling direction. The motion of the units is retrograde, ellipses have maximal size at the surface, and the major semi-axis of the ellipses is also normal to the surface. However, the physical properties of natural solids or structures – such as density and stiffness to tensile stress – vary with depth and distance, so that Rayleigh

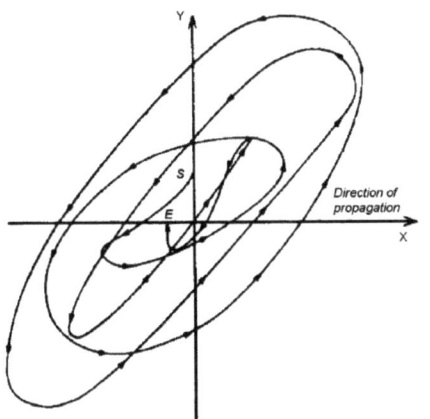

Fig. 12. Irregular motion of a particle on a terrestrial surface during passage of a Rayleigh waves. The waves propagate along the X-axis. S is the particle start position; E is the particle end position. This plot is fragment of Figure 2.13 from Sheriff and Geldart (1982).

waves become dispersed and, in particular, the surface particles experience more complicated motion than that in ellipses (see Figure 12). Since fractured asteroids are certainly non-isotropic bodies, the loosely coupled particles of asteroid regolith will most likely move irregularly under the action of Rayleigh waves, approximately in such a way as shown in Figure 12. If the amplitude of Rayleigh waves is larger than the size of particles, the initial position of a particle and its final position, occurring when the wave passes, will be different (as shown in Figure 12), and so this wave process tends to mix adjacent particles. Since particles shift most in the vertical plane (in relation to the surface plane) the fresh subjacent particles tend to mix with the topmost ones.

The long-term seismic vibrations of asteroid surfaces produce downslope motion of regolith resting before impact on inclined planes (Richardson et al., 2005 and references therein). As a result, fresh unweathered material outcrops due to this process. Richardson et al. (2005) note that this seismically triggered downslope motion is a combination of horizontal sliding and vertical hopping of the upper regolith layer. In other words, under the specified amplitude and frequency of seismic vibrations, a perturbing force predominates over a cohesion force and regolith particles gain one more degree of freedom being revealed as sliding and hopping.

Under the seismic vibrations, the last type of regolith motion is obviously the prevalent event and occurs even if a surface slope angle is zero. One can expect that hopping (or shaking) of regolith particles results in their intermixing and size sorting if they differ in size and/or density (e.g., the reviews of Ottino and Khakhar, 2000 and Kudrolli, 2004 and references therein; Murdoch et al., 2011; Richardson et al., 2011). The larger particles under vertical vibration of granular medium tend to accumulate at the top of the surface; this case is called the Brazil nut effect (e.g. Rosato et al., 1987). However, the reverse Brazil-nut effect may also occur. If the density of large particles is much greater than that of the small ones and the amplitude of vibration is greater than the size of large particles, they begin to move downwards while the small particles rise to the surface (Schröter et al., 2006; Breu et al., 2003).

Features consistent with regolith migration and segregation on the scale of centimeter to several tens meter have been documented on asteroids Eros and Itokawa (Asphaug et al., 2001; Miyamoto et al., 2007). Richardson et al. (2005) have demonstrated the efficacy of seismic erosion and eventual erasing of small impact craters on Eros-sized asteroids. Moreover, the latter is apparently valid for the sub-kilometer-sized asteroid Itokawa in spite of its low density (~1.9 g/cm3), high porosity (~ 40%) and a very likely rubble-pile structure (Abe et al., 2006; Fujiwara et al., 2006). One might expect that this type of asteroid internal structure substantially attenuates seismic energy. Nevertheless, the lack of small craters on Itokawa is consistent with their erasure by seismic shaking due to, in particular, the cumulative effects of many small impacts (Michel et al., 2009).

The seismically induced motion of regolith particles becomes significant only if a dimensionless seismic acceleration ratio $a_s / g > 1$ (g denotes acceleration due to gravity of the given target asteroid). Richardson et al. (2005) have estimated the minimum impactor size, $D_{p,\min} \propto D_a^{5/3}$, necessary to achieve global seismic acceleration equal to g on asteroids of various sizes without their total disruption. Thereby impactors with diameter larger than $D_{p,\min}$ are able to cause global seismic effects throughout the volume of an asteroid, destabilizing all slopes on the surface. It is obvious that each such event greatly reduces space weathering effect for the asteroid surface as a whole due to the extensive surface modification. During the relatively short time span $T_{OC \to S}$, global shaking of surfaces occurs about ~ 1000 and 100 times on asteroids of 20 and 30 km in diameter but is an almost unlikely event if $D_a > 40$ km. This happens because the vast majority of projectiles falling on the asteroids over this time period are small, and their surface-modifying seismic effects are local by their very nature.

It makes sense therefore to estimate the area around impact sites, where seismic acceleration is equal to or more than g. For this, we use the following equation between the seismic energy, E_s, and the distance, r, from the impact site (Yanovskaya, 2008):

$$E_s = \eta E_k = \frac{8\pi^3 A^2}{T} \rho v_p r^2 \exp(2\alpha r). \qquad (17)$$

This equation is written for the first highest-power body wave having the amplitude A and period T and propagating with velocity $v_p.$ Other parameters of Eq. (17) are the following: E_k is the kinetic energy of an impactor, η is the impact seismic efficiency factor, ρ is the mass density of the target asteroid, and α is the attenuation coefficient of the body wave. Keeping in mind that the maximum seismic acceleration, a_s, can be expressed in terms of the maximum displacement A of the medium and frequency $f=1/T$, that is, $a_s = 4\pi^2 f^2 A$, we can rewrite Eq. (16) as follows:

$$r \exp(\alpha r) = a_s^{-1} \left(\frac{2\pi f^3 \eta E_k}{\rho v_p} \right)^{1/2}. \qquad (18)$$

If we set a_s to be equal to g in Eq. (18), then r will equal r_g, the "seismic radius" around the impact site where the seismic and gravity accelerations are equal. Clearly, if $r < r_g$, then $a_s > g$. Note that r (or r_g) is, in fact, the straight-line chord distance between the impact site and a point on asteroid surface (the asteroid body is supposed to be spherical). By analogy with a structure of terrestrial impact craters (Melosh, 1989), we believe also that the range of elastic deformations starts at a distance of $2R_{cr}$ from the point of impact, so that the seismic area around the impact site, where the surface modification processes could occur, is $S_g = \pi(r_g^2 - D_{cr}^2)$.

It is interesting to estimate the total seismic area, S_{tg}, around all impact craters originated on the given asteroid in the time $T_{OC \to S}$. This value scaled to the total asteroid area is

$$S_{tg}/S_a = \frac{1}{D_a^2}\sum_p(r_g^2 - D_{cr}^2)N_p,$$ (19)

where N_p is given by Eq. (3). The numerical values of the parameters required to calculate this ratio were taken from Richardson et al. (2005) and Richardson (2009).

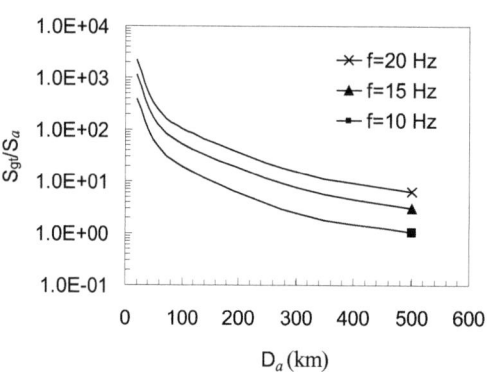

Fig. 13. Total area around an impact site (S_{gt}), where seismic-induced acceleration is greater than or equal to gravitational acceleration on an asteroid's surface as a function of the asteroid's diameter (D_a). These values were estimated for the time interval $T = 1.5\times10^5$ years and scaled to total area of the asteroid surfaces (S_a). f is the primary seismic frequency of surface oscillations (Richardson et al., 2005).

Namely, the mean mass density of asteroids and projectiles, ρ and δ, are assumed to be the same and are equal to 3 g/cm^3; the impact velocity of $U = 5$ km/s; the body wave velocity of $v_p = 0.5$ km/s, i.e., the typical value for the P wave in loose material; $\eta = 10^{-4}$ and $\alpha = 0.069$ km^{-1}. Synthetic seismograms simulating the surface vibrations of fractured asteroid body have a frequency between 1 and 100 Hz with a peak near $10 - 20$ Hz (Richardson et al., 2005). So we used in the calculations three values $f = 10$, 15, and 20 Hz.

Shown in Figure 13 is dependence of the S_{tg}/S_a ratio on asteroid diameters at various values of the surface vibration frequency. As is seen from the Figure, $S_{tg}/S_a > 1$ for asteroids with diameters from 20 to several hundred kilometers. This suggests

that during the time $T_{OC \to S} \sim 1.5\times10^5$ yr every surface element of an asteroid experiences multiple impact excitations with sufficient seismic magnitude to generate the various types of regolith motions discussed above. Specifically, regular mixing of the exposed and subsurface fresh regolith particles appreciably reduces the optical effect of space weathering of the main belt asteroids.

Let us return to Figure 7b in order to consider a correlation between collision resurfacing and surface maturation from another viewpoint. For the same time $T_{OC \to S}$ the 500-km body in the lunar orbit experiences fewer collisions by approximately two orders of magnitude than in the main belt. Thereby if the thickness of ejecta blankets on the body averages ~ 0.5 cm at the distances of the asteroid belt (see Figure 9b), then the value will be less by a factor of 100 at the distance of 1 AU, that is, only 0.005 cm. Since the impact velocity is about 25 km/s at 1 AU, the seismic radii (Eq. 18) are about five times more here than those in the asteroid belt. However, the total seismic area S_{tg} on the 500-km body is nevertheless less by a factor of ~ 4 as against the same value in the asteroid belt and, consequently, $S_{tg}/S_a < 1$ for the body in the lunar orbit. The latter means that the seismic events occur in the immediate vicinity of the impact craters and ~ 75% of the body surface can be accepted as undisturbed. Keeping in mind the trace amount of material deposited for the time $T_{OC \to S}$, one can conclude that the 500-km body at 1 AU maturates according to the lunar type.

4.4. Short and long space weathering timescales

As we have demonstrated, the short weathering timescales inferred from laboratory investigations provide the fast modification of ordinary chondrite spectra to the S-type in ~10^5 years. Then the process stagnates and does not proceed to "completion"; i.e., lunar-type spectra over the next ~ 10^6 yr and beyond. This paradox is especially puzzling since both these time intervals are much shorter than the collisional lifetime (~10^9 yr) of asteroids larger than about 10 km in size (O'Brien and Greenberg, 2005). On the other hand, the theoretical investigations of relationships between the S-asteroid colors, diameters, and ages leads to very long space weathering timescale in the asteroid belt, at least > 10^8 yr (Willman and Jedicke, 2011; 2012).

We have shown above, impact activated motions of regolith particles can hamper the progress of optical maturation of the asteroid surfaces. Moreover, the fact that the space weathering timescale of ~ 2×10^9 yr inferred from S-asteroid spectral observations (Willman and Jedicke, 2011) is much longer than the laboratory estimation $T_{OC \to S}$ based on ion irradiation experiments may be a direct consequence of regular rejuvenation of asteroid surfaces. For this very long time interval, the original (reference) surface of asteroids has been destroyed during the first 10^7 yr, the thickness of the regolith will reach several tens of meters, and global and regional

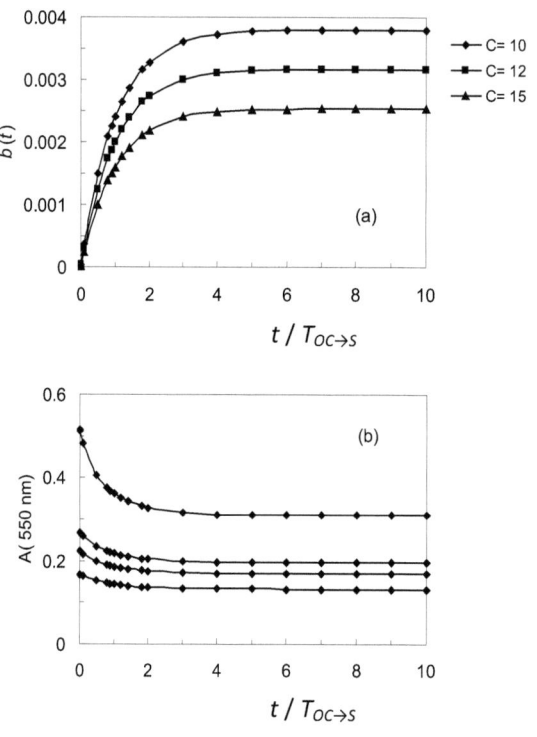

Fig. 14. (a) The time dependence of the maturity degree $b(t)$ of asteroid surfaces according to Eq. (20) at the various values of C parameter, $b_{max} = 0.038$, $T_{OC \to S} = 1.5 \times 10^5$; (b) The time dependence of the visual albedo of asteroids under this maturation process, $C = 10$.

resurfacing events will iteratively happen in the history of asteroids with diameters up to 500 km (Richardson et al., 2005; Richardson, 2011). Therefore we can suppose that the maturation level of asteroid surfaces may be a compromise resulting from a competition between impact resurfacing and solar wind darkening, and after reaching some steady state after a relatively short time (say, several times × $T_{OC \to S}$), no longer depends on time.

If so, the time dependence of the maturity degree (Eq. 2) that we have used for undisturbed surface could be written in the case of the impact-activated asteroid surfaces as follows:

$$b(t) = \frac{b_{max}}{C}\left(1 - \exp(-t/T)\right). \tag{20}$$

Note that Eq. (2) is the special case of Eq. (20) if $t/T \ll 1$ and $C = 1$ for undisturbed and isotropic surface. As before, $b_{max} \approx 0.038$ is the maximal degree of optical maturity when visual albedo of the surface saturates (Fig. 2, *bottom panel*; Fig. 5) and $T = T_{OC \to S}$, the time needed to make any "chondritic" spectrum similar to the S-type. The C parameter cannot be precisely specified within the bounds of our work. We can only roughly estimate $C \sim 10$ or somewhat larger believing reasonably that the maturity degree of non-stationary asteroid regolith should be less than that of an undisturbed surface as the same time $T_{OC \to S}$ passes.

At first, maturity degree rises almost linearly when $t / T_{OC \to S} < 1$ and amounting to some value remains almost constant since $t \sim 5 \times T_{OC \to S}$ (Fig. 14a). The time dependences of asteroid albedo according to Eq. (20) and Shkuratov et al., (1999) theory are shown in Fig. 14b. In contrast to an undisturbed surface, an original asteroid albedo does not saturate up to ~ 0.05 but after approaching some lower value no longer depends on time. It is clear that other optical characteristics of asteroids (such as spectral slopes, color indexes, absorption band depths) also have similar conservative behavior with time if Eq (20) is valid.

So, if the optical maturation of asteroid surfaces happens similar to the above law (Eq. 20) then despite the short weathering timescale occurring for solar wind irradiation, the rate of asteroid surface maturation over a long time could be a very low in accordance with the inference found by Willman and Jedicke (2011).

5. Discussion

Observations of several asteroids *in situ* by spacecraft reveal that the space weathering effect of their surface units correlates with resurfacing processes and differs from lunar type. (McFadden et al., 2001; Bell et al., 2002; Clark et al., 2001, 2003; Chapman, 2004; Ishiguro et al., 2007; Pieters et al., 2012). The spectral contrasts of Eros' dark terrains are believed to be determined by not only lunar-like optical maturation, but also a dark spectrally neutral constituent, presumably shock-modified troilite (Clark et al., 2001; Bell et al., 2002). In turn, spectral analysis of Eros' areas leads to an ordinary chondritic composition (McFadden et al., 2001; Bell et al., 2002), though the freshest bright material shows significant spectral differences from the proposed meteorite analogues (Clark et al., 2001).

Murchie et al. (2002) found a strong correlation between albedo $A(760$ nm) and color $C(950$ nm/760 nm) for spatially resolved areas in the northern hemisphere of asteroid Eros. The authors stressed that space weathering is most likely process responsible for albedo and color differences on Eros though the effects of this process are strongly different from lunar maria. Later on, Shestopalov (2002) has shown that the "Eros line" on the albedo – color plot arises from the following fact: the larger the regolith particles on the asteroid surface, the higher the cumulative amount of reduced iron in these particles. The result quantitatively shows the difference in maturation of the asteroid and lunar regoliths. For the lunar soil, the finest fraction is enriched in reduced iron relative to the large size fraction and dominates optical properties of the lunar bulk material (Hapke, 2001; Noble et al., 2001). The "unusual" behavior of the Eros' regolith particles may be a response to seismic vibrations of the asteroid surface (the Brazil nut effect): the larger the particle, the likelier it is to survive on the surface; the longer its exposure time, the greater the cumulative amount of Fe^0 in the particle.

Unlike asteroid Eros, the surface of Itokawa, the sub-kilometer-sized near-Earth asteroid, shows extensive areas of coarse materials, from centimeter-sized pebbles to boulders of a several tens of meters in size (Miyamoto et al., 2007). Mainly due to this, the Itokawa surface is less mature than that of Ida, Gaspra and, probably, Eros since optical maturation of solids is less effective in comparison with fine-grained regolith (Hapke, 2001; Ishiguro et al., 2007). Maturation level correlates with the surface morphology (Sasaki et al., 2007; Ishiguro et al., 2007) which in turn is a result of various types of granular motions (as landslide-like migrations, particle sorting and segregation and others) triggered mainly by impact-induced seismic vibrations (Miyamoto et al., 2007).

All these undoubtedly significant nuances obvious "at proximate examination" play a minor role in the determination of Itokawa's composition inferred from ground-based spectral observations. The inference that the rocks of Itokawa in whole are similar to L/LL chondrites (Binzel et al., 2001; Abell et al., 2007) was made well before surface specimens were returned by the Hayabusa probe (Nakamura et al., 2011). Moreover, the telescopic investigations of the asteroid (Abell et al., 2007) reveal complex thermal evolution of a parent body before its disruption, one of the pieces of which is known today as Itokawa. Returned samples support the idea of complex regolith evolution, as individual grains have compositions ranging from petrologic grade 3 to 6 and show variable amounts of nanophase iron from grain to grain (Nakamura et al., 2011).

Viewing the asteroid literature one can see that spectroscopic analysis fairly states whether a given S asteroid has chondritic composition (e.g., Binzel et al., 2009) or does not (e.g., Hardersen et al., 2006) regardless of whether the investigators take into account optical effect of space weathering or not. It is not surprising since the maturation level in the asteroid belt is so low that maturation does not appreciably shift band center positions in asteroid spectra; essentially complete vitrification of pre-existing minerals is required for significant alteration of these spectral parameters (Gaffey, 2010).

Figures 1 – 6 show that the b parameter is a measure of the maturation level of a surface. As follows from Eq. (1), if b is sufficiently low, then τ_{Fe} can be much less than τ_h and spectral alteration occurring due to space weathering will be less (or, at least, comparable with) the natural spectral variations of mineral constituents forming the asteroid's surface. This problem becomes relevant when modeling asteroid spectra and not only for the S-types. For example, V asteroids demonstrate low maturation levels, as seen in the best matches between the observed and theoretical spectra (Shestopalov et al., 2008). In the case of S asteroids, the b parameter appears to be less than 0.002, i.e., the value used for undisturbed ordinary chondritic surface (see Figure 2). The numeric simulations associated with the modern physical theories (Shkuratov et al., 1999; Hapke, 2001) are needed to estimate maturation level more accurately for each S-type asteroid studied.

As we have demonstrated, the difference in the degree of optical maturation of the Moon and asteroids results from the unique properties of their regoliths. Unlike the Moon, asteroid regoliths are characterized by higher particle mobility due to the high frequency of non-catastrophic collisions with projectiles and low gravity of asteroid targets. Under such conditions, a thin "optically weathered skin" rich in SMFe steadily wears out due to gardening and does not cover the asteroid surfaces with the same degree of homogeneity as on the Moon. As a consequence, variations of disk-integrated spectral characteristics among S asteroids appear to be caused to a greater extent by variations of surface material composition, particle sizes and porosity of the topmost regolith layer, perhaps vitrification, and agglomeration of regolith particles than their optical maturation under exposure to solar wind.

Unlike our simplifying assumptions (though in the calculations we used the conservative values of problem parameters and did not oversimplify the significance of these and other factors), asteroids differ in composition, internal structure, mechanical and seismic properties. The surface of each asteroid is, in fact, a snapshot of its collisional and proton irradiation histories and the surface maturation level is in its own way a product of the confluence of these processes.

Our results suggest that space weathering processes cannot give rise to color diversity in the main asteroid belt, however we can face the opposite situation in the Edgeworth–Kuiper belt. Ion irradiation experiments simulating solar wind effects on complex carbonaceous compounds and hydrocarbon ices (i.e., potential constituents of the trans-Neptunian objects' and Centaurs' surface materials) also show optical maturation trends, though differing from those for silicate minerals (Moroz et al., 2004). Notably, the probable weathering timescale estimated is short, $\sim 3 \times 10^5$ yr at $\sim$ 40 AU (Moroz et al., 2003). Keeping in mind that the impact velocity here is ~ 0.5 km/s and the collision rate may be less than in the asteroid belt (Davis and Farinella, 1997), space weathering might produce diversity of the optical properties of the Edgeworth–Kuiper belt bodies.

6. Conclusion

Analysis of laboratory experiments simulating space weathering optical effects on atmosphereless planetary bodies reveals that the time needed to alter the spectrum of an ordinary chondrite meteorite to resemble the overall spectral shape and slope of an S-type asteroid is about $\sim 10^5$ yr. The time required to reduce the visible albedo of samples to ~ 0.05 is $\sim 10^6$ yr. Since both these timescales are much less than the average collisional lifetime of asteroids larger than several kilometers in size, numerous low-albedo asteroids having reddish spectra with subdued absorption bands should be observed instead of an S-type dominated population. It is not the case because asteroid surfaces cannot be considered as undisturbed, unlike laboratory samples. We have estimated the number of collisions occurring in the time of 10^5 yr between asteroids and projectiles of various sizes and show that impact-activated motions of regolith particles counteract the progress of optical maturation of asteroid surfaces. Continual rejuvenation of asteroid surfaces by impacts does not allow bodies with the ordinary chondrite composition to be masked among S asteroids. Spectroscopic analysis, using relatively invariant spectral parameters, such as band centers and band area ratios, can determine whether the surface of an S asteroid has chondritic composition or not. Differences in the environment of the main asteroid belt versus that at 1 AU, and the physical difference between the Moon and main belt asteroids (i.e., size) can account for the lack of lunar-type weathering on main belt asteroids.

The theoretical model that could combine regolith generation, retention, and movement on asteroid surfaces together with their resulting optical effects under solar wind is still in development (e. g., Vernazza et al., 2009; Marchi et al., 2012). In practice, the following conclusions remain valid:

(i) The low degree of asteroid surface maturity does not impede the remote mineralogical analysis of asteroids by their reflectance spectra.

(ii) The optical maturation of asteroid surfaces should remain under consideration when simulating asteroid spectra; the scattering theories (Shkuratov et al., 1999; Hapke, 2001) have been successfully applied to evaluate the stage of optical maturity and composition of lunar and asteroid regoliths.

References

Abe, S., Mukai, T., Hirata, N., Barnouin-Jha, O.S., Cheng, A.F., Demura, H., Gaskell, R.W., Hashimoto, T., Hiraoka, K., Hona, T., Kubota, T., Matsuoka, M., Mizuno, T., Nakamura, R., Scheeres, D.J., Yoshikawa, M., 2006. Mass and local topography measurements of Itokawa by Hayabusa. Science 312, 1344–1347.

Abell, P. A., Vilas, F., Jarvis, K. S., Gaffey, M. J., Kelley, M. S., 2007. Mineralogical composition of (25143) Itokawa 1998 SF$_{36}$ from visible and near-infrared reflectance spectroscopy: Evidence for partial melting. Meteorit. Planet. Sci. 42, 2165–2177.

Asphaug, E., King, P.J., Swift, M.R., Merrifield, M.R., 2001. Brazil nuts on Eros: Size-sorting of asteroid regolith. Planet. Sci. Conf. XXXII., 1708.

Bell, J. F. III, Izenberg, N. I., Lucey, P. G., Clark, B. E., Peterson, C., Gaffey, M. J., Joseph, J., Carcich, B., Harch, A., Bell, M. E., Warren, J., Martin, P. D., McFadden, L. A., Wellnitz, D., Murchie, S., Winter, M., Veverka, J., Thomas, P., Robinson, M. S., Malin, M., Cheng, A., 2002. Near-IR reflectance spectroscopy of 433 Eros from the NIS instrument on the NEAR mission. 1. Low phase angle observations. Icarus 155, 119–144.

Binzel, R. P., Rivkin, A. S., Bus, S. J., Sunshine, J. M., Burbine, T. H., 2001. MUSES-C target asteroid (25143) 1998 SF$_{36}$: A reddened ordinary chondrite. Meteorit. Planet. Sci. 36, 1167–1172.

Binzel, R. P., Rivkin, A. S., Thomas, C. A., Vernazza, P., Burbine, T. H., DeMeo, F. E., Bus, S J., Tokunaga, A. T., Birlan, M., 2009. Spectral properties and composition of potentially hazardous asteroid (99942) Apophis. Icarus 200, 480–485.

Binzel, R. P., Morbidelli A., Merouane, S., DeMeo, F. E., Birlan, M., Vernazza, P., Thomas, C. A., Rivkin, A. S., Bus, S. J., Tokunaga, A. T., 2010a. Good vibrations: Recent near-earth encounters as the missing piece of the S-asteroid and ordinary chondrite meteorite puzzle. Lunar and Planet. Sci. Conf. IVI. Abstr. #1226.

Binzel, R. P., Morbidelli, A., Merouane, S., DeMeo, F. E., Birlan, M., Vernazza, P., Thomas, C. A., Rivkin, A. S., Bus, S. J., Tokunaga, A. T., 2010b. Earth encounters as the origin of fresh surfaces on near-Earth asteroids. Nature 463, 331–334.

Borg, J., Comstock G. M., Langevin, Y., Maurette, M., Jouffrey, B., Jouret, C., 1976. A Monte Carlo model for the exposure history of lunar dust grains in the ancient solar wind. Earth Planet. Sci. Lett. 29, 161–174.

Bottke, W.F., Greenberg, R., 1993. Asteroidal collision probabilities. Geophys. Res. Lett. 20, 879–881.

Breu, A. P. J., Ensner, H.-M., Kruelle, C. A., Rehberg, I., 2003. Reversing the Brazil-Nut Effect: Competition between Percolation and Condensation. Phys. Rev. Lett. 90, 014302, 3 pages.

Brunetto, R., Strazzulla, G., 2005. Elastic collisions in ion irradiation experiments: A mechanism for space weathering of silicates. Icarus 179, 265–273.

Brunetto, R., Vernazza, P., Marchi, S., Birlan, M., Fulchignoni, M., Orofino, V., Strazzulla, G., 2006a. Modeling asteroid surfaces from observations and irradiation experiments: The case of 832 Karin. Icarus 184, 327–337.

Brunetto, R., Romano, F., Blanco, A., Fonti, S., Martino, M., Orofino, V., Verrienti, C., 2006b. Space weathering of silicates simulated by nanosecond pulse UV excimer laser. Icarus 180, 546–554.

Cellino, A., Delbò, M., Bendjoya, Ph., Tedesco, E. F., 2010. Polarimetric evidence of close similarity between members of the Karin and Koronis dynamical families. Icarus 209, 556–563.

Chapman, C. R., 1996. S-type asteroids, ordinary chondrites, and space weathering: The evidence from *Galileo's* fly-bys of Gaspra and Ida. Meteorit. Planet. Sci. 31, 699–725.

Chapman, C. R., 2004. Space weathering of asteroid surfaces. Annu. Rev. Earth Planet. Sci. 32, 539–567.

Clark, B. E., Lucey, P., Helfenstein, P., Bell, J. F. III, Peterson, C., Veverka, J., McConnochie, T., Robinson, M., Bussey, B., Murchie, S., Izenberg, N., Chapman, C., 2001. Space weathering on Eros: Constraints from albedo and spectral measurements of Psyche crater. Meteorit. Planet. Sci. 36, 1617–1637.

Clark, B. E., Hapke, B., Pieters, C., Britt, D., 2003. Asteroid space weathering and regolith evolution. In: Bottke, W. F., Jr., Cellino, A., Paolicchi, P., Binzel, R.P. (Eds.), Asteroids III. Univ. Arizona Press, pp. 585–599.

Cloutis, E. A., Klima, R. L., Kaletzke, L., Coradini, A., Golubeva, L. F., McFadden, L. A., Shestopalov, D. I., Vilas, F., 2010. The 506 nm absorption feature in pyroxene spectra: Nature and implications for spectroscopy-based studies of pyroxene-bearing targets. Icarus 207, 295–313.

Dainty, M.A., Toksöz, M.N., Anderson, K.R., Pines, P.J., Nakamura, Y., Latham, G., 1974. Seismic scattering and shallow structure of the Moon in Oceanus Procellarum. The Moon 9, 11–29.

Davis, D. R., Farinella, P., 1997. Collisional evolution of Edgeworth–Kuiper belt objects. Icarus 125, 50–60.

Davis, D. R., Durda, D. D., Marzari, F., Campo Bagatin A., Gil-Hutton, R., 2003. Collisional Evolution of Small-Body Populations. In: Bottke, W. F., Jr., Cellino, A., Paolicchi, P., Binzel, R.P. (Eds.), Asteroids III. Univ. Arizona Press, pp. 545–558.

Dukes, C., Baragiola, R. A., McFadden, L. A., 1999. Surface Modification of Olivine by H+ and He+ Bombardment, J. Geophys. Res. 104, 1865–1872.

Farinella, P, Paolicchi, P., Zappala, V., 1982. The asteroids as outcomes of catastrophic collisions. Icarus 52, 409–433.

Fujiwara, A., and 21 colleagues, 2006. The near-Earth Asteroid Itokawa observed by Hayabusa: Its possible structure and history. Science 312, 1330–1334.

Gaffey, M.J., 1976. Spectral reflectance characteristics of meteorite classes. J. Geophys. Res. 81, 905–920.

Gaffey, M.J., Bell, J.F., Brown, R.H., Burbine, T.H., Piatek, J.L., Reed, K.L., Chaky, D.A., 1993. Mineralogical variations within the S-type asteroid class. Icarus 106, 573–602.

Gaffey, M. J., 2010. Space weathering and the interpretation of asteroid reflectance spectra. Icarus 209, 564–574.

Golubeva, L. F., Shestopalov, D. I., 1992. S-asteroids of the main belt and meteorite sources. Solar System Res. 26, 602–608.

Golubeva, L. F., Shestopalov, D. I., 2002. On the origin of the last meteorite parent bodies. Solar System Res. 36, 478–486.

Golubeva, L.F., Shestopalov, D.I., 2003. Albedo (750 nm) – color (950/750 nm) diagram for the Moon, asteroids and meteorites: Modeling of optical maturation of the cosmic body surfaces. Lunar and Planet. Sci. Conf. XXXIV. Abstr. #1096.

Golubeva, L. F., Shestopalov, D. I., 2011. Space weathering of asteroids. Lunar and Planet. Sci. Conf. IVII. Abstr. #1029.

Golubeva, L. F., Shestopalov, D. I., Shkuratov Yu. G., 1980. A comparative analysis of certain optical characteristics of asteroids and the Moon. Sov. Astron. 24, 600–605.

Hapke, B., 1973. Darkening of silicate rock powders by solar wind spattering. The Moon 7, 342–355.

Hapke, B., 2001. Space weathering from Mercury to asteroid belt. J. Geophys. Res. 106, NO. E5, 10039– 10073.

Hardersen, P. S., Gaffey, M. J., Cloutis, E. A., Abell, P. A., Reddy, V., 2006. Near-infrared spectral observations and interpretations for S asteroids 138 Tolosa, 306 Unitas, 346 Hermentaria, and 480 Hansa. Icarus 181, 94–106.

Hiroi, T., Vilas, F., Sunshine, J.M., 1996. Discovery and analysis of minor absorption bands in S-asteroid visible reflectance spectra. Icarus 119, 202–208.

Holsapple, K. A., 1993. The scaling of impact processes in planetary sciences. Annu. Rev. Earth Planet. Sci. 21, 333–373.

Hörz, F., Cintala, M. J., See, T. H., Le, L., 2005. Shock melting of ordinary chondrite powders and implications for asteroidal regoliths. Meteorit. Planet. Sci. 40, 1329–1346.

Housen, K. R., Holsapple, K. A., 2011. Ejecta from impact craters. Icarus 211, 856–875.

Housen, K. R., Schmidt, R. M., Holsapple, K. A., 1983. Crater ejecta scaling laws: Fundamental forms based on dimensional analysis. J. Geophys. Res. 88, 2485–2499.

Housen, K. R., Wilkening, L. L., 1982. Regoliths on small bodies in the Solar System. Ann. Rev. Earth Planet. Sci. 10, 355–376.

Housen, K. R., Wilkening, L. L., Chapman, C. R., Greenberg, R., 1979. Asteroidal regolith. Icarus 39, 317–351.

Ishiguro, M., Hiroi, T., Tholen, D. J., Sasaki, S., Ueda, Y., Nimura, T., Abe, M., Clark, B. E., Yamamoto A., Yoshida, F., Nakamura, R., Hirata, N., Miyamoto, Yokota, H., Y., Hashimoto, T., Kubota, T., Nakamura, A. M., Gaskell, R. W., Saito, J., 2007. Global mapping of the degree of space weathering on asteroid 25143 Itokawa by Hayabusa/AMICA observations. Meteorit. Planet. Sci. 42, 1791–1800.

Johnson, P. B., Cristy, R. W., 1974. Optical constants of metals: Ti, V, Cr, Mn, Fe, Co, Ni, and Pd. Phys. Rev. B. 9, 5056–5070.

Keller, L., McKay, D., 1997. The nature and origin of rims on lunar soil grains. Geochim. Cosmochim. Acta 61, 2331–2340.

Kudrolli, A., 2004. Size separation in vibrated granular matter. Reports on Progress in Physics, 67, 209–247.

Lazzarin, M., Marchi, S., Moroz, L., V., Brunetto, R., Magrin, S., Paolicchi P., Strazzulla G., 2006. Space weathering in the main asteroid belt: The big picture. Astrophys. J., 647, L179–L182.

Loeffler, M.J., Baragiola, R.A., Murayama, M., 2008. Laboratory simulations of redeposition of impact ejecta on mineral surfaces. Icarus 196, 285–292.

Loeffler, M.J., Dukes, C.A., Baragiola, R.A., 2009. Irradiation of olivine by 4 keV He$^+$: Simulation of space weathering by the solar wind. J. Geophys. Res. – Planets 114, E03003. doi:10.1029/2008JE003249.

Lucey, P. G., Blewett, D. T., Hawke B. R., 1998. Mapping of FeO and TiO_2 content of the lunar surface with multispectral imagery. J. Geophys. Res. 103, 3679–3699.

Lucey, P. G., Riner, M. A., 2011. The optical effects of small iron particles that darken but do not redden: Evidence of intense space weathering on Mercury. Icarus 212, 451–462.

Marchi, S., Magrin, S., Nesvorný, D., Paolicchi, P., Lazzarin, M., 2006a. A spectral slope versus perihelion distance correlation for planet-crossing asteroids. Mon. Not. R. Astron. Soc. 368, L39–L42.

Marchi, S., Paolicchi, P., Lazzarin, M., Magrin, S., 2006b. A general spectral slope–exposure relation for S-type main belt and near-earth asteroids. Astron. J. 131, 1138–1141,

Marchi, S., Paolicchi, P., Richardson, D. C., 2012. I. Collisional evolution and reddening of asteroid surfaces: The problem of conflicting timescales and the role of size-dependent effects. Mon. Not. R. Astron. Soc. 421, 2–8.

Matson, D. L., Johnson, T. V., Veeder, G. J., 1977. Soil maturity and planetary regolith: The Moon, Mercury and asteroids. Proc. Lunar Sci. Conf. 8th, 625–627.

McBride, N. and Hamilton, D. 2000. Meteoroids at high altitudes, final report of ESA contract 13145/98/nl/wk. Technical report, European Space Agency.

McFadden, L. A., Gaffey, M. J., McCord, T. B., 1985. Near-Earth asteroids: Possible sources from reflectance spectroscopy. Science 229, 160–163.

McFadden, L. A., Wellnitz, D.D., Schnaubelt, M., Gaffey, M.J., Bell III, J. F., Izenberg, N., Murchie, S., Chapman, C.R., 2001. Mineralogical Interpretation of Reflectance Spectra of Eros from NEAR near-infrared spectrometer low phase flyby. Meteorit. Planet. Sci. 36, 1711–1726.

Melosh H., 1989. Impact Cratering: A Geologic Process. Oxford University Press, New York, 245 p.

Michel, P., O'Brien, D. P., Abe, S., Hirata, N., 2009. Itokawa's cratering record as observed by Hayabusa: Implications for its age and collisional history. Icarus 200, 503–513.

Miyamoto, H., Yano, H., Scheeres, D. J., Abe, S., Barnouin-Jha, O., Cheng, A. F., Demura, H., Gaskell, R., W., Hirata, N., Ishiguro, M., Michikami, T., Nakamura, A. M., Nakamura, R., Saito, J., Sasaki, S., 2007. Regolith migration and sorting on asteroid Itokawa. Science 316, 1011–1014.

Moroz, L.V., Fisenko, A.V., Semjonova, L.F., Pieters, C.M., Korotaeva, N.N., 1996. Optical effects of regolith processes on S-asteroids as simulated by laser shots on ordinary chondrite and other mafic materials. Icarus 122, 366–382.

Moroz, L. V., Baratta, G., Distefano, E., Strazzulla,. G., Starukhina, L. V., Dotto, E., Barucci M. A., 2003. Ion irradiation of asphaltite: Optical effects and implications for trans-Neptunian objects and Centaurs. Earth, Moon and Planets 92, 279–289.

Moroz, L., Baratta, G., Strazzulla, G., Starukhina, L., Dotto, E., Barucci, M. A., Arnold, G., Distefano, E., 2004, Optical alteration of complex organics induced by ion irradiation: 1. Laboratory experiments suggest unusual space weathering trend. Icarus 170, 214–228.

Morris, R. V., 1977. Surface and near-surface exposure ages of lunar soils: exposure ages based on the fine-grained metal and cosmogenic [21]Ne. Abstr. Lunar and Planet. Sci. Conf. VIII, 685–687.

Morris, R. V., 1980. Origins and size distribution of metallic iron particles in the lunar regolith. Proc. Lunar Planet. Sci. Conf. XI. 1697–1712.

Murchie, S., Robinson, M., Clark, B., Li, H., Thomas, P., Joseph, J., Bussey, B., Domingue, D., Veverka, J., Izenberg, N., Chapman, C., 2002. Color variations on Eros from NEAR multispectral imaging. Icarus, 155, 145–168.

Mothé-Diniz, T., Jasmin, F.L., Carvano, J.M., Lazzaro, D., Nesvorný, D., Ramirez, A.C., 2010. Re-assessing the ordinary chondrites paradox. Astron. Astroph. 514, A86.

Murdoch, N., Michel, P., Richardson, D. C., Walsh, K. J., Losert, W., Berardi, C., Green, S. F. 2011. Numerical Simulations of Granular Dynamics in Various Conditions Applicable to Regolith Motion on Small Body Surfaces. Planet. Sci. Conf. XXXXII. Abstr. #1113.

Nakamura, T., Noguchi, T., Tanaka, M., Zolensky, M. E., Kimura, M., Tsuchiyama, A., Nakato, A., Ogami, T., Ishida, H., Uesugi, M., Yada, T., Shirai, K., Fujimura, A., Okazaki, R., Sandford, S. A., Ishibashi, Y., Abe, M., Okada, T., Ueno, M., Mukai, T., Yoshikawa, M., Kawaguchi, J., 2011. Itokawa dust particles: a direct link between S-type asteroids and ordinary chondrites. Science 333, 1113–1116.

Nesvorný, D., Jedicke, R., Whiteley, R. J., Ivezić, Ž., 2005. Evidence for asteroid space weathering from the Sloan Digital Sky Survey. Icarus 173, 132–152.

Nesvorný, D., Vokrouhlický, D., Morbidelli, A., Bottke, W. F., 2009. Asteroidal source of L chondrite meteorites. Icarus 200, 698–701.

Nesvorný, D., Bottke, W. F., Vokrouhlický, D., Chapman, C. R., Rafkin, S., 2010. Do planetary encounters reset surfaces of near Earth asteroids? Icarus 209, 510–519.

Noble, S. K., Pieters, C. M., Taylor, L. A., Morris, R. V., Allen, C. C., McKay, D. S., Keller, L. P., 2001. The optical properties of the finest fraction of lunar soil: Implication for space weathering. Meteorit. Planet. Sci. 36, 31–42.

Noble, S. K., Pieters, C. M., Keller L. P., 2007. An experimental approach to understanding the optical effects of space weathering. Icarus 192, 629–642.

O'Brien, D. P., Greenberg, R., 2005. The collisional and dynamical evolution of the main-belt and NEA size distributions. Icarus 178, 179–212.

O'Brien, D. P., Sykes, M. V., Tricarico, P., 2011. Collision probabilities and impact velocity distributions for Vesta and Ceres. Lunar Planet. Sci. Conf. IVII. Abstr. #2665.

O'Keefe, J. D., Ahrens, T. J., 1977. Impact-induced energy partitioning, melting, and vaporization on terrestrial planets. Proc. Lunar Planet. Sci. Conf. VIII. 3357–3374.

Öpik, E. J., 1951. Collisional probabilities with the planets and the distribution of interplanetary matter. Proc. Royal Irish Acad. 54A, 164–199.

Ottino, J. M., Khakhar, D. V., 2000. Mixing and segregation of granular materials. Annu. Rev. Fluid Mech. 32, 55–91.

Paolicchi, P., Marchi, S., Nesvorný, D., Magrin, S., Lazzarin, M., 2007. Towards a general model of space weathering of S-complex asteroids and ordinary chondrites. Astron. Astroph. 464, 1139–1146.

Paolicchi, P., Marchi, S., Lazzarin, M., Magrin, S., 2009. Collisional timing of asteroids space weathering: A first approach. Planet. Space Sci. 57, 216–220

Pieters, C. M., Taylor, L. A., Noble, S. K., Keller, L. P., Hapke B., Morris, R. V., Allen, C. C., McKay, D. S., Wentworth, S., 2000. Space weathering on airless bodies: Resolving a mystery with lunar samples. Meteorit. Planet. Sci. 35, 1101–1107.

Pieters, C. M., Ammannito, E., Blewett, D. T., Denevi, B. W., de Sanctis, M. C., Gaffey, M. J., Le Corre, L., Li, J.-Y., Marchi, S., McCord, T. B., McFadden, L. A., Mittlefehldt, D. W., Nathues, A., Palmer, E., Reddy, V., Raymond, C. A., Russell, C. T., 2012. Distinctive space weathering on Vesta from regolith mixing processes. Nature 491, pp. 79 – 82.

Reddy, V., Nathues, A., Le Corre, L., Sierks, H., Li, J., Gaskell, R., McCoy, T., Beck, A. W., Schröder, S. E., Pieters, C. M., and 17 coauthors, 2012. Color and Albedo Heterogeneity of Vesta from Dawn. Science 336, 700–704.

Richardson, D. C., Walsh, K. J., Murdoch, N., Michel, P., 2011. Numerical simulations of granular dynamics: I. Hard-sphere discrete element method and tests. Icarus 212, 427–437.

Richardson Jr, J. E., Melosh, H. J., Greenberg , R. J., O'Brien, D. P., 2005. The global effects of impact-induced seismic activity on fractured asteroid surface morphology. Icarus 179, 325–349.

Richardson, J. E., Melosh, H. J., Lisse, C. M., Carcich, B., 2007. A ballistics analysis of the Deep Impact ejecta plume: Determining Comet Tempel 1's gravity, mass, and density. Icarus 190, 357–390.

Richardson, J. E., 2009. Cratering saturation and equilibrium: A new model looks at an old problem. Icarus 204, 697–715.

Richardson, J. E., 2011. Regolith generation, retention, and movement on asteroid surfaces: early modeling results. Planet. Sci. Conf. XXXXII. Abstr. #1084.

Rivkin, A.S., Thomas, C.A., Trilling, D.E., Enga, M., Grier, J.A., 2011. Ordinary chondrite-colors in small Koronis family members. Icarus 211, 1294–1297.

Rosato, A., Strandburg, K. J., Prinz, F., Swendsen, R. H., 1987. Why the Brazil nuts are on top: Size segregation of particulate matter by shaking. Phys. Rev. Lett. 58, 1038–1040.

Sanchez, J. A., Reddy, V., Nathues, A., Cloutis, E. A, Mann, P., Hiesinger, H., 2012. Phase reddening on near-Earth asteroids: Implications for mineralogical analysis, space weathering and taxonomic classification. Icarus 220, 36 – 50.

Sasaki, S., Nakamura, K., Hamabe, Y., Kurahashi, E,. Hiroi, T., 2001. Production of iron nanoparticles by laser irradiation in a simulation of lunar-like space weathering. Nature 410, 555–557.

Sasaki, S., Ishiguro, M., Hirata, N., Hiroi, T., Abe, M., Abe, S., Miyamoto, H., Saito, J., Yamamoto, A., Demura, H., Kitazato, K., Nakamura, R., 2007. Origin of surface albedo/color variation on rubble-pile Itokawa. Lunar Planet. Sci. Conf. XXXVIII. Abstr. #1293.

Schenk, P., O'Brien, D. P., Marchi, S., Gaskell, R., Preusker, F., Roatsch, T., Jaumann, R., Buczkowski, D., McCord, T., McSween, H. Y., and 4 coauthors, 2012. The Geologically Recent Giant Impact Basins at Vesta's South Pole. Science 336, 694–697.

Schröter, M., Ulrich, S., Kreft, J., Swift, J. B., Swinney, H. L., 2006. Mechanisms in the size segregation of a binary granular mixture. arXiv:cond-mat/0601179v3 [cond-mat.soft].

Sheriff, R. E., Geldart, L. P., 1982. Exploration seismology. Volume 1: History, theory, & data acquisition. 447 p. Cambridge Univ. Press of Cambridge, England.

Shestopalov, D., 2002. About optical maturation of Eros regolith. Proceed. of Asteroids, Comets, Meteors. Berlin, Germany (ESA-SP-500, November 2002), 919–921.

Shestopalov, D. I., Golubeva L. F., 2004. The optical maturation of a chondrite surface: Modeling for S-type asteroids and meteorites. Solar System Res., 38, 203–211.

Shestopalov, D. I., Golubeva, L. F., 2008. Why Vesta's surface is unweathered? Lunar Planet. Sci. Conf. XXXIX. Abstr. #1116.

Shestopalov, D., Sasaki, S., 2003. Calculations of optical effects of the laser experiment imitating space weathering of the cosmic body surfaces. Lunar Planet. Sci. Conf. XXXIV, Abstract # 1097.

Shestopalov, D.I., McFadden, L.A., Golubeva, L.F., 2007. Exploration of faint absorption bands in the reflectance spectra of the asteroids by method of optimal smoothing: Vestoids. Icarus 187, 469–481.

Shestopalov, D.I., Golubeva, L.F., Taran, M.N., Khomenko, V.M., 1991. Iron and chromium absorption bands in the spectra of terrestrial pyroxenes: Application to mineralogical remote sensing of asteroid surfaces. Solar System Res. 25, 442–452.

Shestopalov, D.I., McFadden, L.A., Golubeva, L.F., Khomenko, V.M., Gasanova, L.O., 2008. Vestoid surface composition from analysis of faint absorption bands in visible reflectance spectra. Icarus 195, 649–662.

Shkuratov, Yu., Starukhina, L., Hoffman, H., and Arnold, G., 1999. A model of spectral albedo of particulate surfaces: implications to optical properties of the moon, Icarus 137, 235–246.

Starukhina, L. V., 2003. Computer simulation of sputtering of lunar regolith by solar wind protons: Contribution to change of surface composition and to hydrogen flux at the lunar poles. Solar Sys. Res. 37, 36–50.

Starukhina, L. V., Shkuratov Yu. G., 2001. A theoretical model of lunar optical maturation: Effects of submicroscopic reduced iron and particle size variations. Icarus 152, 275–281.

Starukhina, L. V., Shkuratov, Y. G., 2011. Reduced iron grains from nano- to micron sizes in lunar and mercurian regoliths: calculation of spectral effects. Lunar and Planet. Sci. Conf. IVII. Abstract # 1144.

Strazzulla, G., Dotto, E., Binzel, R., Brunetto, R., Barucci, M.A., Blanco, A., Orofino, V., 2005. Spectral alteration of the meteorite Epinal (H5) induced by heavy ion irradiation: A simulation of space weathering effects on near-Earth asteroids. Icarus 174, 31–35.

Sullivan, R.J., Thomas, P.C., Murchie, S.L., Robinson, M.S., 2002. Asteroid geology from Galileo and Near Shoemaker data. In: Bottke, W.F., Jr., Cellino, A., Paolicchi, P., Binzel, R.P. (Eds.), Asteroids III. Univ. Arizona Press, pp. 331–350.

Thomas, C.A., Rivkin, A.S., Trilling, D.E., Enga, M., Grier, J.A., 2011. Space weathering of small Koronis family members. Icarus 212, 158–166.

Thomas, C. A., Trilling, D. E., Rivkin, A. S., 2012. Space weathering of small Koronis family asteroids in the SDSS Moving Object Catalog. Icarus 219, 505–507.

Toksöz, M.N., 1975. Lunar and Planetary Seismology. Rev. Geophys. Space Phys. 13, 306–311.

Vernazza, P., Binzel, R., Birlan, Fulchignoni, M., Rossi, A., 2009. Solar wind as the origin of rapid reddening of asteroid surfaces. Nature 458, 993–995.

Wentworth, S., Keller, L., McKay, D., Morris, R., 1999. Space weathering on the Moon: Patina on Apollo 17 samples 75075 and 76015. Meteorit. Planet. Sci. 34, 593–603.

Werner, S. C., Harris, A. W., Neukum, G., 2002. The Near-Earth Asteroid Size–Frequency Distribution: A Snapshot of the Lunar Impactor Size–Frequency Distribution. Icarus 156, 287–290.

Wetherill, G. W., 1967. Collisions in the asteroid belt. J. Geophys. Res. 72, 2429–2444.

Willman, M., Jedicke, R., Nesvorný D., Moskovitz, N., Ivezić, Ž., Fevig, R., 2008., Redetermination of the space weathering rate using spectra of Iannini asteroid family members. Icarus 195, 663–673.

Willman, M., Jedicke, R., Moskovitz, N., Nesvorny´, D., Vokrouhlicky´, D., Mothé-Diniz, T., 2010. Using the youngest asteroid clusters to constrain the space weathering rate on S-complex asteroids. Icarus 208, 758–772.

Willman, M., Jedicke, R., 2011. Asteroid age distributions determined by space weathering and collisional evolution models. Icarus 211, 504–510.

Willman, M., Jedicke, R., 2012. Corrigendum to "Asteroid age distributions determined by space weathering and collisional evolution models" [Icarus 211 (2011) 504–510]. Icarus 217, 431.

Yamada, M., Sasaki, S., Nagahara, H., Fujiwara, A., Hasegawa, S., Yano, H., Hiroi T., Ohashi, H., Otake, H., 1999. Simulation of space weathering of planet forming materials: Nanosecond pulse laser irradiation and proton implantation on olivine and pyroxene samples. Earth Planets Space 51, 1255–1265.

Yanovskaya, T. B., 2008. Fundamentals of seismology. Univ. S.-Petersburg Press, Russia, 260 p.